ENERGY SCIENCE, ENGINEERING AND TECHNOLOGY

WIND ENERGY

FEDERAL INITIATIVES AND EFFECTIVE FINANCIAL SUPPORT

Energy Science, Engineering and Technology

Additional books in this series can be found on Nova's website under the Series tab.

Additional E-books in this series can be found on Nova's website under the E-book tab.

Energy Policies, Politics and Prices

Additional books in this series can be found on Nova's website under the Series tab.

Additional E-books in this series can be found on Nova's website under the E-book tab.

ENERGY SCIENCE, ENGINEERING AND TECHNOLOGY

WIND ENERGY

FEDERAL INITIATIVES AND EFFECTIVE FINANCIAL SUPPORT

ANAS DESMET
EDITOR

nova publishers
New York

For permission to use material from this book please contact us:
Telephone 631-231-7269; Fax 631-231-8175
Web Site: http://www.novapublishers.com

Library of Congress Cataloging-in-Publication Data

ISBN: 978-1-62808-069-8

Published by Nova Science Publishers, Inc. † New York

CONTENTS

PREFACE

Wind energy has been the fastest growing source of U.S. electric power generation in recent years. The increase in federal funding for wind technologies and involvement of multiple agencies has raised questions about fragmented, overlapping, or duplicative federal support. This book examines federal wind-related initiatives; programs or groups of agency activities that promoted wind energy through a specific emphasis or focus. The Government Accountability Office (GAO) identifies wind-related initiatives implemented by federal agencies in fiscal year 2011; assesses the extent of fragmentation, overlap, and duplication, if any, among these initiatives, and the extent to which they were coordinated; and examines how agencies allocate support to projects through their initiatives and the extent to which they assess applicant need for support.

Chapter 1 – Wind energy has been the fastest growing source of U.S. electric power generation in recent years. The increase in federal funding for wind technologies and involvement of multiple agencies has raised questions about fragmented, overlapping, or duplicative federal support.

In this report, GAO examines federal wind-related initiatives—programs or groups of agency activities that promoted wind energy through a specific emphasis or focus. GAO (1) identifies wind-related initiatives implemented by federal agencies in fiscal year 2011 and their key characteristics; (2) assesses the extent of fragmentation, overlap, and duplication, if any, among these initiatives, and the extent to which they were coordinated; and (3) examines how agencies allocate support to projects through their initiatives and the extent to which they assess applicant need for support. GAO sent a questionnaire to agencies to identify wind-related initiatives and to obtain data on their characteristics; potential for fragmentation, overlap, or duplication;

and related coordination. GAO also reviewed studies of the initiatives and interviewed agency officials and financial professionals.

Chapter 2 – Increasing U.S. energy supply diversity has been the goal of many Presidents and Congresses. This commitment has been prompted by concerns about national security, the environment, and the U.S. balance of payments. Investments in new energy sources also have been seen as a way to expand domestic manufacturing. For all of these reasons, the federal government has a variety of policies to promote wind power.

Expanding the use of wind energy requires installation of wind turbines. These are complex machines composed of some 8,000 components, created from basic industrial materials such as steel, aluminum, concrete, and fiberglass. Major components in a wind turbine include the rotor blades, a nacelle and controls (the heart and brain of a wind turbine), a tower, and other parts such as large bearings, transformers, gearboxes, and generators. Turbine manufacturing involves an extensive supply chain. Until recently, Europe has been the hub for turbine production, supported by national renewable energy deployment policies in countries such as Denmark, Germany, and Spain. However, support for renewable energy including wind power has begun to wane across Europe as governments there reduce or remove some subsidies. Competitive wind turbine manufacturing sectors are also located in India and Japan and are emerging in China and South Korea.

U.S. and foreign manufacturers have expanded their capacity in the United States to assemble and produce wind turbines and components. About 470 U.S. manufacturing facilities produced wind turbines and components in 2011, up from as few as 30 in 2004. An estimated 30,000 U.S. workers were employed in the manufacturing of wind turbines in 2011. Because turbine blades, towers, and certain other components are large and difficult to transport, manufacturing clusters have developed in certain states, notably Colorado, Iowa, and Texas, which offer proximity to the best locations for wind energy production. The U.S. wind turbine manufacturing industry also depends on imports, with the majority coming from European countries, where the technical ability to produce large wind turbines was developed. Although turbine manufacturers' supply chains are global, recent investments are estimated to have raised the share of parts manufactured in the United States to 67% in 2011, up from 35% in 2005-2006.

The outlook for wind turbine manufacturing in the United States is more uncertain now than in recent years. For the past two decades, a variety of federal laws and state policies have encouraged both wind energy production and the use of U.S.-made equipment to generate that energy. A continuing challenge for the industry is uncertainty about one main federal policy tool in the deployment of wind power, the production tax credit (PTC), which Congress has extended eight times and let lapse on four occasions. Most recently, the PTC expired at the end of 2012, but a few days later, Congress extended it through year-end 2013. At least a dozen wind turbine manufacturers announced layoffs or hiring freezes at U.S. facilities in 2012, citing concern about the PTC's future as one reason. Other factors affecting the health of the U.S. wind industry are intense price competition from natural gas, an oversupply in wind turbines, and softening demand for renewable electricity.

In: Wind Energy
Editor: Anas Desmet
ISBN: 978-1-62808-069-8

Chapter 1

WIND ENERGY: ADDITIONAL ACTIONS COULD HELP ENSURE EFFECTIVE USE OF FEDERAL FINANCIAL SUPPORT*

United States Government Accountability Office

WHY GAO DID THIS STUDY

Wind energy has been the fastest growing source of U.S. electric power generation in recent years. The increase in federal funding for wind technologies and involvement of multiple agencies has raised questions about fragmented, overlapping, or duplicative federal support.

In this report, GAO examines federal wind-related initiatives—programs or groups of agency activities that promoted wind energy through a specific emphasis or focus. GAO (1) identifies wind-related initiatives implemented by federal agencies in fiscal year 2011 and their key characteristics; (2) assesses the extent of fragmentation, overlap, and duplication, if any, among these initiatives, and the extent to which they were coordinated; and (3) examines how agencies allocate support to projects through their initiatives and the extent to which they assess applicant need for support. GAO sent a questionnaire to agencies to identify wind-related initiatives and to obtain data on their characteristics; potential for fragmentation, overlap, or duplication;

* This is an edited, reformatted and augmented version of United States Government Accountability Office, Publication No. GAO-13-136, dated March 2013.

and related coordination. GAO also reviewed studies of the initiatives and interviewed agency officials and financial professionals.

What GAO Recommends

GAO recommends that to the extent possible within their statutory authority DOE and USDA formally assess and document whether the federal financial support of their initiatives is needed for applicants' wind projects to be built. DOE agreed with the recommendation and USDA generally concurred with the findings related to its initiatives.

What GAO Found

GAO identified 82 federal wind-related initiatives, with a variety of key characteristics, implemented by nine agencies in fiscal year 2011. Five agencies—the Departments of Energy (DOE), the Interior, Agriculture (USDA), Commerce, and the Treasury—collectively implemented 73 of the initiatives. The 82 initiatives incurred about $2.9 billion in wind-related obligations and provided estimated wind-related tax subsidies totaling at least $1.1 billion in fiscal year 2011, although complete data on wind-related tax subsidies were not available. Initiatives supporting deployment of wind facilities, such as those financing their construction or use, constituted the majority of initiatives and accounted for nearly all obligations and estimated tax subsidies related to wind in fiscal year 2011. In particular, a tax expenditure and a grant initiative, both administered by Treasury, accounted for nearly all federal financial support for wind energy.

The 82 wind-related initiatives GAO identified were fragmented across agencies, most had overlapping characteristics, and several that financed deployment of wind facilities provided some duplicative financial support. The 82 initiatives were fragmented because they were implemented across nine agencies, and 68 overlapped with at least one other initiative because of shared characteristics. About half of all initiatives reported formal coordination. Such coordination can, in principle, reduce the risk of unnecessary duplication and improve the effectiveness of federal efforts. However, GAO identified 7 initiatives that have provided duplicative support—financial support from multiple initiatives to the same recipient for deployment of a single project.

Specifically, wind project developers have in many cases combined the support of more than 1 Treasury initiative and, in some cases, have received additional support from smaller grant or loan guarantee programs at DOE or USDA. GAO also identified 3 other initiatives that did not fund any wind projects in fiscal year 2011 but that could, based on their eligibility criteria, be combined with 1 or more initiatives to provide duplicative support. Of the 10 initiatives, those at Treasury accounted for over 95 percent of the federal financial support for wind in fiscal year 2011.

Agencies implementing the 10 initiatives allocate support to projects on the basis of the initiatives' goals or eligibility criteria, but the extent to which applicant financial need is considered is unclear. DOE and USDA—which have some discretion over the projects they support through their initiatives—allocate support based on projects' ability to meet initiative goals such as reducing emissions or benefitting rural communities, as well as other criteria. Both agencies also consider applicant need for the support of some initiatives, according to officials.

However, GAO found that neither agency documents assessments of applicant need; therefore the extent to which they use such assessments to determine how much support to provide is unclear. Unlike DOE and USDA, Treasury generally supports projects based on the tax code's eligibility criteria and does not have discretion to allocate support to projects based on need. While the support of these initiatives may be necessary in many cases for wind projects to be built, because agencies do not document assessments of need, it is unclear, in some cases, if the entire amount of federal support provided was necessary. Federal support in excess of what is needed to induce projects to be built could instead be used to induce other projects to be built or simply withheld, thereby reducing federal expenditures.

ABBREVIATIONS

accelerated depreciation	Accelerated Depreciation Recovery Periods for Specific Energy Property
BOEMRE	Bureau of Ocean Energy Management, Regulation and Enforcement
Commerce	Department of Commerce
DOE	Department of Energy
DSIRE	Database of State Incentives for Renewables and Efficiency

EERE	Office of Energy Efficiency and Renewable Energy
EIA	Energy Information Administration
EPA	Environmental Protection Agency
FERC	Federal Energy Regulatory Commission
Interior	Department of the Interior
IRS	Internal Revenue Service
ITC	Energy Investment Credit, also known as the Investment Tax Credit
LBNL	Lawrence Berkeley National Laboratory
NSF	National Science Foundation
PTC	Energy Production Credit, also known as the Production Tax Credit
REAP	Rural Energy for America Program
Recovery Act	American Recovery and Reinvestment Act of 2009
RPS	Renewable Portfolio Standard
SBA	Small Business Administration
Section 1603 program	Payments for Specific Energy Property in Lieu of Tax Credits
Section 1703 program	Title XVII Section 1703 Loan Guarantee Program
Section 1705 program	Title XVII Section 1705 Loan Guarantee Program
Treasury	Department of the Treasury
USDA	U.S. Department of Agriculture

March 11, 2013
The Honorable Lamar Smith
Chairman
Committee on Science, Space, and Technology
House of Representatives

The Honorable Cynthia M. Lummis
Chairman
Subcommittee on Energy
Committee on Science, Space, and Technology
House of Representatives

The Honorable Paul Broun, M.D.
Chairman
Subcommittee on Oversight
Committee on Science, Space, and Technology
House of Representatives

Americans' daily lives, as well as the economic productivity of the United States, depend on the availability of energy, the majority of which comes from fossil fuels. However, concerns over the nation's reliance on imported oil, volatile energy costs, and fossil fuels' emissions of greenhouse gases linked to global climate change have increased the focus on developing renewable energy resources to help meet future energy needs. One of these resources—wind energy—has been the fastest growing source of U.S. electric power generation in recent years, increasing at about 33 percent per year since 2001, according to the Department of Energy's (DOE) Energy Information Administration (EIA), an independent statistical and analytical agency. In 2011, wind energy comprised 32 percent of all new additions to U.S. electric generating capacity and contributed 3.0 percent of our nation's total electricity generation, the largest share of any nonhydroelectric renewable source. While helping meet the nation's energy needs, wind energy may offer substantial environmental benefits as well, such as a reduction in greenhouse gas emissions, compared with traditional energy sources. However, wind energy as a source of electricity also faces numerous challenges related to its intermittent availability and historically higher costs compared with traditional energy sources.

In support of efforts to research, develop, and deploy wind energy technologies, the federal government provided nearly $5 billion in subsidies in fiscal year 2010—more than 75 percent of federal subsidies for all renewable sources of electricity, according to EIA estimates.[1] Of this amount, about $4.9 billion was related to the American Recovery and Reinvestment Act of 2009 (Recovery Act).[2] By contrast, in fiscal year 2007—the previous year for which EIA analyzed these data—EIA estimated the federal government provided $476 million to support wind energy technologies.[3] This funding has helped spur development of wind energy in the United States, helping to meet goals (e.g., for energy security) established by federal agencies and policymakers. DOE has a leading role in committing federal resources and expertise to research, develop, and deploy wind energy technologies, as part of its mission to address the nation's energy and environmental challenges through science and technology. The Department of the Treasury (Treasury), meanwhile, provides large tax subsidies for wind energy technologies, through its administration of provisions in the federal Internal Revenue Code.[4] Other agencies play a role in supporting wind energy technologies as well; in February 2012, we reported that 23 federal agencies had implemented nearly 700 renewable energy initiatives—programs or groups of agency activities that involved renewable energy through a specific emphasis or focus—in fiscal year 2010.[5] Many of these initiatives supported wind energy technologies. In addition, states also play an important role in encouraging development of wind and other renewable energy sources through policies and financial incentives. For example, many states have renewable portfolio standards (RPS) that require public utilities to generate or buy a percentage of energy from renewable sources.

In the current fiscally constrained environment, effective allocation of resources is especially important. The increase in funding for wind energy technologies and the involvement of multiple agencies have raised questions that federal efforts to support these technologies may be fragmented, overlapping, or duplicative. Fragmentation occurs when more than one federal agency, or more than one organization within an agency, is involved in furthering the development of wind energy. For purposes of this report, overlap occurs when multiple initiatives support similar wind issues,[6] similar technology advancement activities,[7] similar recipients, and have similar goals. As we have previously reported, fragmentation and overlap among government programs can potentially lead to unnecessary duplication.[8] For purposes of this report, duplication occurs when multiple initiatives provide financial support to the same recipient for a single wind project. Duplication as we have defined it may be necessary, in some cases, for specific wind projects to be built. However, in other cases, duplication may result in ineffective use of federal financial support—that is, it may result in some amount of support being provided for specific wind projects that is not needed for them to be built.

In this report, we examine federal initiatives that promoted wind energy, which we termed "wind-related initiatives."[9] Our objectives for this report were to (1) identify wind-related initiatives implemented by federal agencies in fiscal year 2011 and their key characteristics; (2) assess the extent of fragmentation, overlap, and duplication, if any, among these initiatives, and the extent to which they were coordinated; and (3) examine how agencies allocate support to projects through their initiatives and the extent to which they assess applicant need for support.

To address these objectives, we collected and analyzed information from our previous work and conducted new work. Our review focused on those initiatives that promoted the research and development, commercialization, or deployment of wind energy technologies in fiscal year 2011, the most recent year for which data were available. To identify these initiatives, we relied, in part, on data from our February 2012 report on renewable energy initiatives, which identified and collected information from wind- and other renewable energy-related initiatives active in fiscal year 2010. Using these data, we focused our review on research and development, commercialization, or deployment initiatives for which wind energy was eligible for support. In addition, we excluded certain agencies—such as the Departments of Defense, Homeland Security, and State—which have initiatives generally focused on development of wind energy and other technologies for use in a military, border security, or international aid setting, rather than for use in the domestic commercial energy market.

For the remaining agencies, we developed a questionnaire to collect information from their officials regarding whether their fiscal year 2010 initiatives were still active and whether wind energy still received or was eligible for support under the initiatives in fiscal year 2011. We also asked agency officials to identify any additional initiatives that were active and for which wind energy was eligible for support in fiscal year 2011. Using the responses from this questionnaire, we identified 82 wind-related initiatives at nine agencies. We then developed a second questionnaire and administered it to agency officials to collect additional data on the characteristics of these initiatives; potential fragmentation, overlap, or duplication among them; and efforts to coordinate the initiatives. Specifically, we asked questions regarding initiatives' obligations, or revenue losses from tax expenditures;[10] year in which the initiatives first supported wind energy; type of wind issues, technology advancement activities, and recipients supported; initiative goals; and efforts to coordinate with other wind-related initiatives. We received survey responses from all 82 initiatives and therefore had a response rate of 100 percent. (For further information on our objectives, scope, and methodology, including the use of our questionnaire, see app. I. For a copy of our questionnaire, see app. IV.) We then analyzed the data collected from the questionnaire to categorize initiatives' recipients and goals, and to determine the extent of potential fragmentation, overlap, and duplication of wind-related initiatives. For selected initiatives, we collected information from agency websites on financial support provided for projects, and we interviewed agency officials for information on their agencies' efforts to assess applicants' need for the support of their initiatives. We also reviewed studies of the initiatives by DOE's national laboratories, the Congressional Research Service, and other experts. We interviewed financial professionals from several of the major financial institutions and legal firms active in wind energy project financing in recent years. To obtain additional information about the types of support available to wind project developers, we collected and analyzed data from the Database of State Incentives for Renewables and Efficiency (DSIRE), a comprehensive source of information on state incentives and policies that promote renewable energy and energy efficiency, which is funded by DOE. We interviewed researchers who developed and maintain DSIRE and determined the data were sufficiently reliable for our purposes. We also interviewed agency officials and financial professionals for additional information on state initiatives.

We conducted this performance audit from February 2012 to March 2013 in accordance with generally accepted government auditing standards. Those standards require that we plan and perform the audit to obtain sufficient, appropriate evidence to provide a reasonable basis for our findings and conclusions based on our audit objectives. We believe that the evidence obtained provides a reasonable basis for our findings and conclusions based on our audit objectives.

BACKGROUND

Wind energy is generated when wind turbines convert the kinetic energy in the wind into mechanical power, which can be used to generate electricity or for specific tasks such as grinding grain or pumping water. Electricity generated from wind can be used for other stand-alone purposes, such as charging a battery, or can flow to consumers from the facilities where it is generated through the networks that carry electricity, including wires, substations, and transformers (i.e., the grid). For utility-scale sources of wind energy, a large number of turbines are usually built close together to form a wind farm that provides grid power.[11] Stand-alone or distributed turbines are generally smaller scale turbines used for purposes such as powering communications equipment or generating electricity for local use by farmers. Areas with plentiful wind resources are often distant from consumer markets, and access to transmission is essential to bring electricity generated from wind to market. In addition, because of the intermittent availability of wind energy, integrating increasing amounts of wind energy into the electric grid while maintaining its reliable operation requires added efforts by federal agencies, electric grid operators, utilities, and regulators. Another important issue in the development of wind projects is the siting and permitting process, by which locations are chosen and permits are issued for wind turbines, while considering projects' potential effects on the environment and the competing uses for the land, airspace, or waterways the projects may require.

Innovation in wind energy technology takes place across a spectrum of activities, which we refer to as technology advancement activities, and which include basic research, applied research, demonstration, commercialization, and deployment.[12] For purposes of this report, basic research includes efforts to explore and define scientific or engineering concepts or to investigate the nature of a subject without targeting any specific technology; applied research includes efforts to develop new scientific or engineering knowledge to create new and improved technologies; demonstration includes efforts to operate new or improved technologies to collect information on their performance and assess readiness for widespread use; commercialization includes efforts to transition technologies to commercial applications by bridging the gap between research and demonstration activities and venture capital funding and marketing activities; and deployment includes efforts that facilitate or achieve widespread use of technologies in the commercial market.

Wind energy technology advancement activities are financed through both public and private investment. According to a Congressional Budget Office report,[13] without public investment, the private sector's investment in technology advancement activities is likely to be inefficiently low from society's perspective because firms cannot easily capture the "spillover benefits" that result, particularly at the early stages of developing a technology. In these stages, technology advancement activities can create fundamental knowledge leading to numerous benefits for society as a whole but not necessarily for the firms that invested in the activities. For example, basic research can create general scientific knowledge that is not itself subject to commercialization but that can lead to multiple applications that private companies can produce and sell. As activities get closer to the commercialization and deployment stages, the private sector may increase its support because its return on investment is likely to increase.

Federal investment and policies can have a significant impact on wind development. For example, a key tax incentive for the construction of wind projects—the Energy Production Credit (also known as the Production Tax Credit, or PTC)—has periodically expired and then been extended. In years following its expiration, new additions of wind energy capacity fell dramatically, as shown in figure 1 below.

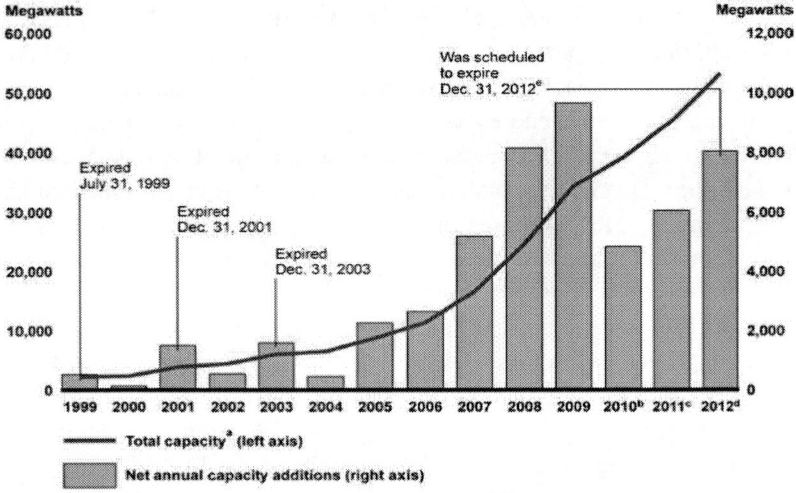

Source: GAO analysis of DOE data.

[a] These data reflect net summer capacity—the maximum output that generating equipment can supply during the period of summer peak demand from June 1 through September 30.

[b] There was no expiration of the PTC immediately prior to 2010; however, the economic downturn and an uncertain regulatory environment (particularly relating to the renewal of production and investment tax credits) contributed to a lower level of wind capacity additions in 2010, according to EIA.

[c] Data for 2011 are preliminary. Final data were not available as of November 2012.

[d] Data for 2012 are projected. Actual data were not available as of November 2012. New wind energy capacity installations may be considerably larger in 2012 than this projection shows. For example, the American Wind Energy Association reports that 4,728 MW of wind energy were installed through the first three quarters of 2012, and there were an additional 8,430 MW under construction as of the third quarter of 2012. In addition, according to DOE, new capacity additions in 2012 were anticipated to exceed 2011 levels and perhaps even 2009 levels as developers rushed to commission projects before the scheduled expiration of several federal tax incentives for wind energy at the end of 2012.

[e] The PTC for wind projects was extended on January 2, 2013, by the American Taxpayer Relief Act of 2012 for facilities that begin construction by December 31, 2013.

Figure 1. New Wind Energy Capacity Additions and Production Tax Credit Expirations, 1999 through 2012.

Many states have also enacted policies affecting the development of wind energy, in part to attract investment within their borders. These policies include tax credits, grants, loans, and mandates such as RPSs requiring a portion of the electricity consumed or generated in a state to be from renewable sources. Improvements in technology and external market factors also affect the development of wind energy. According to a report from DOE's Office of Energy Efficiency and Renewable Energy (EERE), recent improvements in the cost and performance of wind energy technologies contributed to the growth of wind energy in 2011.[14] For example, improvements such as taller towers and larger rotor diameters in wind turbines have improved their efficiency. However, according to the EERE report, continued low natural gas prices and modest growth in electricity demand, among other factors, may dramatically slow new installations of wind turbines in 2013. In addition, wind energy must compete in the market with other energy sources—renewable and nonrenewable—that are also receiving subsidies. For example, in a 2011 report, EIA estimated that the federal government provided nearly $6.7 billion in subsidies for coal, natural gas, petroleum liquids, and nuclear energy in fiscal year 2010.[15]

Multiple Federal Agencies Promoted Wind Energy in Fiscal Year 2011 through Numerous Initiatives Mainly Supporting Deployment

We identified 82 federal wind-related initiatives, with a variety of key characteristics, implemented by nine agencies in fiscal year 2011.[16] Five agencies—DOE, the Department of the Interior (Interior), the Department of Agriculture (USDA), the Department of Commerce (Commerce), and Treasury—collectively implemented 73 of these initiatives. In fiscal year 2011, wind-related initiatives incurred about $2.9 billion in obligations for activities specifically related to wind. In addition to initiatives that obligated funds, Treasury's wind-related tax expenditure initiatives provided estimated tax subsidies of at least $1.1 billion for activities specifically related to wind, although complete data on wind-related tax subsidies were not available.[17] The initiatives supported a range of wind issues, including siting and permitting, offshore wind, and, most commonly, utility-scale and distributed land-based wind. They also supported a range of technology advancement activities, from basic and applied research to, most commonly, deployment. The majority of initiatives provided funding or other direct support for energy providers, developers, or manufacturers, and less than half of the initiatives supported other types of recipients such as public and private researchers or individuals. Initiatives supporting deployment accounted for all tax subsidies and nearly all obligations related to wind in fiscal year 2011. In particular, a tax expenditure and a grant initiative, both at Treasury, accounted for nearly all federal financial support related to wind.

Five Agencies Implemented Nearly All Initiatives and Were Responsible for Several Billion Dollars in Federal Support Related to Wind

Of the nine agencies that implemented the 82 federal wind-related initiatives we identified in fiscal year 2011, five lead agencies—DOE, Interior, USDA, Commerce, and Treasury—were collectively responsible for 73 (89 percent) of the initiatives. (See app. II for a full list of the initiatives.) The remaining four agencies—the Environmental Protection Agency (EPA), the Federal Energy Regulatory Commission (FERC), the National Science Foundation (NSF), and the Small Business Administration (SBA)—each had three or fewer initiatives. Figure 2 shows the number of initiatives by agency.

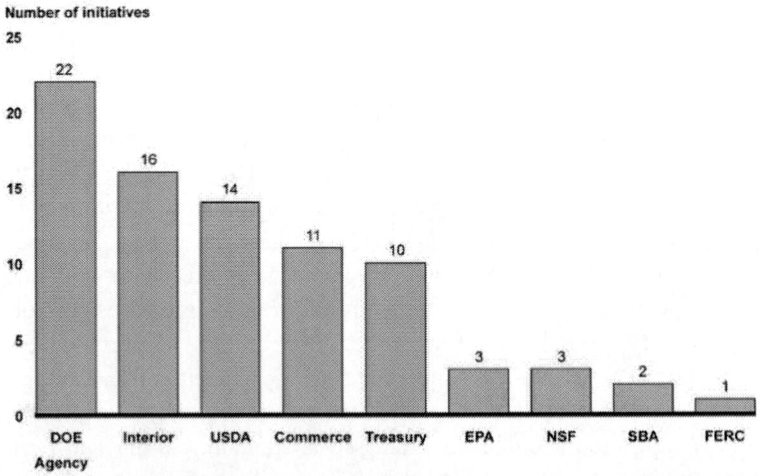

Source: GAO analysis of agency-provided data.

Figure 2. Number of Federal Wind-Related Initiatives by Agency, in Fiscal Year 2011.

Around half of the initiatives—43 of 82—began supporting wind energy in fiscal year 2008 or before. For instance, the PTC, which provides an income tax credit based on the amount of energy produced at wind and other qualified facilities, was first enacted under the Energy Policy Act of 1992. However, several key initiatives began supporting wind energy more recently as part of the Recovery Act or other recent legislation. For instance, Section 1603 of the Recovery Act created Treasury's Payments for Specific Energy Property in Lieu of Tax Credits initiative (Section 1603 program), which provides cash payments of up to 30 percent of the total eligible costs of wind and certain other renewable energy facilities, in lieu of tax credits for energy investment or production. Key initiatives such as these and the PTC for wind facilities have recently expired or are scheduled to expire at the end of 2013. (See app. II for a full list of initiatives, including information on their expiration dates.)

The majority of the wind-related initiatives we identified supported a range of renewable energy sources in addition to wind, as well as other activities such as energy efficiency projects or rural development projects. Specifically, 16 initiatives (20 percent) supported wind energy either exclusively or primarily, and 51 initiatives (62 percent) supported other renewable energy sources or other activities either primarily or equally with wind energy.[18] For instance, initiatives that exclusively or primarily supported wind energy included several under EERE's Wind Energy Program that focused on research, development, and testing to improve the performance, lower the costs, and accelerate the deployment of wind technologies, and several at Interior's Bureau of Ocean Energy Management, Regulation, and Enforcement (BOEMRE) that focused on facilitating the development of offshore wind through resource assessments, environmental impact studies, and granting of leases and rights-of-way for projects.[19] In contrast, wind-related tax expenditures such as the PTC, as well as several initiatives enacted or expanded under the Recovery Act such as Treasury's Section 1603 program, supported a range of renewable energy sources and, in some cases, other sources such as nuclear energy; energy efficiency projects related to buildings or vehicles; or carbon capture and storage projects involving coal or other fossil fuel sources.[20]

In responding to our questionnaire, agency officials reported that they obligated around $2.9 billion through their initiatives in fiscal year 2011 for activities specifically related to wind.[21] These obligations data represent a mix of actual obligations and estimates.[22]

Table 1. Actual and Estimated Obligations for Activities Specifically Related to Wind, by Agency, in Fiscal Year 2011

Agency[a]	Number of wind-related initiatives for which data were provided	Actual obligations	Estimated obligations	Total obligations
Treasury	1	$2,716,933,281	$0	$2,716,933,281
DOE	17	73,940,581	73,161,968	147,102,549[b]
Interior	15	10,206,170	15,778,339	25,984,509
USDA	9	4,850,539	56,000	4,906,539[c]
Commerce	4	2,332,038	415,462	2,747,500
NSF	2	2,104,544	0	2,104,544
EPA	2	30,000	210,000	240,000
Total	50	$2,810,397,153	$89,621,769	$2,900,018,922

Source: GAO analysis of agency-provided data.

Note: Because we summarize obligations data by agency, agency-level data typically reflect a mix of actual and estimated obligations. However, obligations reported for any specific initiative are either actual or estimated. For instance, EPA's data on its two initiatives above reflect one initiative for which actual obligations of $30,000 were reported, and one for which estimated obligations of $210,000 were reported.

[a] In addition to the 50 initiatives at the seven agencies listed here, FERC did not provide obligations data for its one wind-related initiative because it noted that all costs to the government associated with the initiative are recovered through charges to regulated entities. SBA did not provide obligations data for either of its two initiatives because, according to agency officials, one initiative provided loan guarantees whose costs were offset by fees, and the second initiative was in the early planning stages in fiscal year 2011.

[b] Of the $147 million obligated by DOE for activities specifically related to wind in 2011, about $51 million was obligated for credit subsidy costs—the government's estimated net long-term cost, in present value terms, of the loans it guarantees as part of the Title XVII Section 1705 Loan Guarantee Program. Credit subsidy costs exclude administrative costs and any incidental effects on governmental receipts or outlays. Present value is the worth of the future stream of returns or costs in terms of money paid immediately. In calculating present value, prevailing interest rates provide the basis for converting future amounts into their current equivalents.

[c] This amount does not reflect a guarantee for a $204 million loan provided for a wind project in fiscal year 2011 through USDA's Direct and Guaranteed Electric Loan Program. USDA officials said that based on the historical performance of the loans and the creditworthiness of applicants for the program, they estimate zero credit subsidy costs for the program.

Officials for 39 initiatives (about 48 percent) reported actual obligations, officials for 13 initiatives (about 16 percent) reported estimated obligations, and officials for 21 initiatives (about 26 percent) were not able to provide estimated or actual obligations.[23]

Officials who provided estimates or were unable to provide obligations data noted that the accuracy or the availability of the obligations data was limited because, for example, isolating the obligations for activities specifically related to wind can be difficult.

In addition, among the 21 initiatives for which no wind-specific obligations data were reported, agency officials for several of them reported, for example, that they recovered their costs from power customers, or they provided loan guarantees whose costs were offset by fees paid by lenders. As shown in table 1, Treasury was responsible for nearly 94 percent of the total reported obligations (about $2.7 billion of $2.9 billion), all of which were due to its Section 1603 program.

In addition, Treasury's nine other wind-related initiatives were tax expenditures that provided estimated tax subsidies totaling at least $1.1 billion for activities specifically related to wind, according to available estimates. This amount is based almost solely on subsidies provided through the PTC, which was the only one of these nine initiatives for which complete estimates of wind-specific tax subsidies were available for 2011.[24]

Both Treasury and the Joint Committee on Taxation develop estimates of tax subsidies provided through these tax expenditure initiatives; however, the initiatives support a range of renewable and other energy sources, and wind-specific estimates are not available for most of them. (See app. III for a list of these tax expenditures and the available estimates of revenue losses related to wind energy.)

Initiatives Supported a Range of Wind Issues, Technology Advancement Activities, and Recipients

The 82 initiatives we identified supported a range of issues related to wind, and most initiatives supported more than one wind issue. The wind issues most commonly supported were utility-scale land-based wind (49 initiatives) and distributed land-based wind (45 initiatives). See table 2 below for the number and percentage of initiatives supporting each wind issue.

Table 2. Number and Percentage of Federal Wind-Related Initiatives Supporting Each Wind Issue in Fiscal Year 2011

Wind issue	Number of initiatives	Percentage of initiatives
Utility-scale land-based wind	49	60
Distributed land-based wind	45	55
Offshore wind	38	46
Grid integration	37	45
Transmission	33	40
Siting and permitting	28	34

Source: GAO analysis of agency-provided data.

Note: Because many wind-related initiatives supported multiple wind issues, the percentage of initiatives does not total 100, and the number of initiatives does not total 82, the number of initiatives identified in our review.

Individual initiatives tended to support a range of wind issues. Specifically, 63 initiatives (77 percent) supported more than one wind issue, and 46 of these (56 percent of all initiatives) supported three or more wind issues.[25] For instance, USDA's High Energy Cost Grant Program provides grants for energy facilities and infrastructure serving rural communities with average home energy costs exceeding 275 percent of the national average. Officials responding to our questionnaire reported that these grants can support utility-scale and distributed land-based wind, transmission, and grid integration. Similarly, DOE's State Energy Program provides financial and technical assistance to state governments for a variety of renewable energy-related activities across all six wind issues, according to agency officials.

The 82 initiatives we identified also supported technology advancement activities ranging from basic and applied research through deployment, which was the most commonly supported technology advancement activity; 58 initiatives (71 percent) included support for deployment. See table 3 below for the number and percentage of initiatives supporting each technology advancement activity. Our analysis showed that 39 (48 percent) of the 82 initiatives supported only one type of technology advancement activity. Another 39 initiatives (48 percent) supported more than one type of technology advancement activity, and of these, 5 initiatives supported all five.[26] For example, Commerce's Joint Wind Energy Program: Atmospheric Velocity Gradients initiative supports a single technology advancement activity—applied research—through studies focused on improving predictions of wind energy production from winds at various heights. In contrast, Commerce's Green Technology Pilot Program supported all types of technology advancement activities except basic research. The program supported investment in wind and other "green" technologies through expedited reviews of patent applications, allowing for earlier intellectual property protection.

Table 3. Number and Percentage of Federal Wind-Related Initiatives Supporting Each Technology Advancement Activity in Fiscal Year 2011

Technology advancement activity	Number of initiatives	Percentage of initiatives
Basic research	13	16
Applied research	34	41
Demonstration	29	35
Commercialization	24	29
Deployment	58	71

Source: GAO analysis of agency-provided data.

Note: Because many wind-related initiatives supported multiple technology advancement activities, the percentage of initiatives does not total 100, and the number of initiatives does not total 82, the number of initiatives identified in our review.

The majority of initiatives supported recipients, generally in the private sector, that provide electricity generated from wind to consumers, develop wind energy generation projects, or manufacture wind-related equipment, according to agency officials responding to our questionnaire. Specifically, 57 initiatives (70 percent) provided funding or other direct support to energy providers, developers, or manufacturers. Fewer than half of the initiatives provided funding or other direct support to recipients such as public and private researchers (35 initiatives), states and other governmental organizations (34 initiatives), and individuals (12 initiatives). Around half of the initiatives—44 initiatives (54 percent)—supported one type of recipient, while the remaining 38 initiatives supported multiple types of recipients.

Initiatives Supporting Deployment Accounted for Most Obligations and Tax Subsidies Related to Wind

In terms of federal financial support, deployment was the primary focus of federal efforts to promote wind energy. Of the reported $2.9 billion in actual and estimated obligations for wind-related activities in fiscal year 2011, $2.86

billion (99 percent) was obligated by the 58 initiatives that included support for deployment. As previously noted, approximately 94 percent of total wind-related obligations—just over $2.7 billion—was obligated by Treasury's Section 1603 program for the deployment of projects. Other initiatives that supported deployment activities obligated $147 million. In addition to obligations, all nine of Treasury's wind-related tax expenditures—with estimated tax subsidies of at least $1.1 billion for activities specifically related to wind—included support for deployment. These tax subsidies were primarily provided through the PTC, which, in 2011, provided an income tax credit of 2.2 cents per kilowatt hour for energy produced from wind and certain other renewable energy sources.[27] In addition, deployment was the most commonly supported technology advancement activity at the five lead agencies.[28] See figure 3 below for the number of initiatives supporting each technology advancement activity at these agencies.

In addition to initiatives that support deployment of wind energy technologies by directly funding or providing tax subsidies for the construction or operation of wind facilities, some initiatives that we identified supported the deployment of these technologies indirectly. This indirect support includes facilitating the buying and selling of wind technologies or wind energy, or encouraging deployment through policies and regulations. Examples include the following:

- Commerce's International Trade Administration implements an International Buyer Program, which supports U.S. companies at trade shows—including several major shows focused on renewable energy—by recruiting and escorting foreign buyer delegations to meet with U.S. companies.
- EPA's Green Power Partnership supports deployment of wind and other renewable energy technologies by encouraging organizations and individuals to purchase renewable energy through outreach, education, and technical support.
- DOE's Division of Permitting, Siting, and Analysis implements an initiative providing technical and financial assistance to state and regional entities, such as public utility commissions and state legislatures, to help develop renewable energy policies and portfolio standards, among other things.

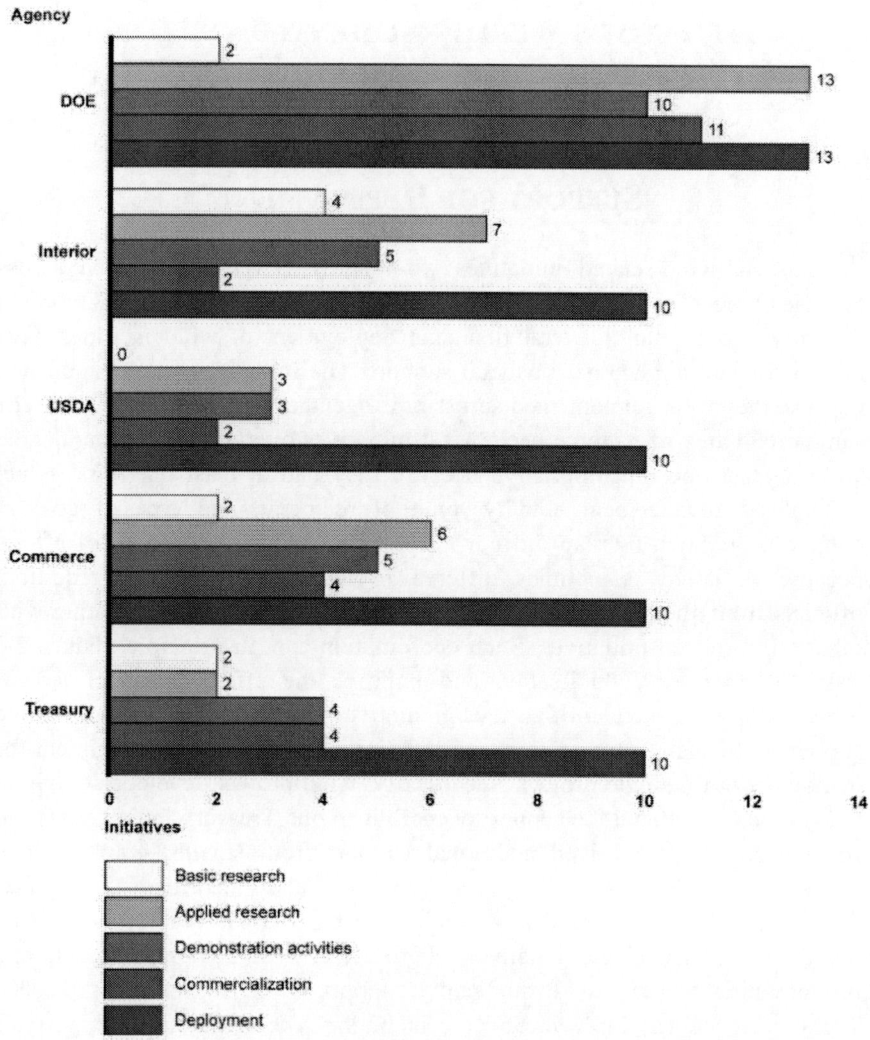

Source: GAO analysis of agency-provided data.

Note: Agency initiatives often supported more than one technology advancement activity; therefore, the sum of initiatives shown in this figure for a given agency will not total the number of that agency's initiatives in our review.

Figure 3. Number of Federal Wind-Related Initiatives at Five Lead Agencies, by Agency and Technology Advancement Activity Supported.

Initiatives Were Fragmented and Had Overlapping Characteristics with Half Reporting Coordination, but Several Provided Some Duplicative Financial Support for Deployment

The 82 wind-related initiatives we identified were fragmented across agencies, most had overlapping characteristics and, though half reported formally coordinating, several financing deployment of wind facilities have provided some duplicative financial support. The initiatives were fragmented because they were implemented across nine agencies and were involved in the same broad area of national need. Most initiatives overlapped to some degree with at least one other initiative because they had at least one wind issue, technology advancement activity, type of recipient, and type of goal in common, but such overlap did not necessarily lead to duplication of efforts because initiatives sometimes differed in meaningful ways. In addition, officials from about half of all initiatives reported formally coordinating with other wind-related initiatives. Such coordination can, in principle, reduce the risk of unnecessary duplication and improve the effectiveness of federal efforts. However, we identified seven initiatives that have provided duplicative support— financial support from multiple initiatives to the same recipient for deployment of a single project. Specifically, wind project developers have, in many cases, combined the support of more than one Treasury initiative and, in some cases, have received additional support from smaller grant or loan guarantee programs at DOE or USDA. We also identified three other initiatives that did not fund any wind projects in fiscal year 2011 but that could, on the basis of the initiatives' eligibility criteria, be combined with one or more initiatives to provide duplicative support. Of the 10 initiatives, those at Treasury accounted for over 95 percent of the federal financial support for wind in fiscal year 2011.

Initiatives Were Fragmented and Most Had Broadly Overlapping Characteristics

The 82 wind-related initiatives we identified were fragmented because they were implemented across nine agencies and were involved in the same broad area of national need: promoting or enabling the development of wind energy. We found that initiatives supporting deployment in particular were spread about evenly across the five lead agencies—each had between 10 and 13 initiatives that supported deployment. In March 2011, we reported that fragmentation has the potential to result in duplication of resources.[29] However, such fragmentation is, by itself, not an indication that unnecessary duplication of efforts or activities exists. For example, in our March 2011 report, we stated that there can be advantages to having multiple federal agencies involved in a broad area of national need—agencies can tailor initiatives to suit their specific missions and needs, among other things.

Across all initiatives, we found that 68 (83 percent) overlapped to some degree with at least one other initiative because they supported similar wind issues, technology advancement activities, and recipients, and had similar goals. The following are several examples of overlapping initiatives:

- *Deployment of utility-scale land-based wind facilities by the energy industry*. Seventeen initiatives provided financial support for the construction or use of utility-scale land-based wind facilities to energy companies. For instance, USDA's Direct and Guaranteed Electric Loan Program provides loans and loan guarantees to establish and improve electric service in rural areas, including through utility-scale on-grid wind and other renewable energy systems. Similarly, the SBA's Certified Development Company/Section 504 Loans initiative guarantees SBA loans to businesses for, among other things, energy efficiency and renewable energy projects, including utility-scale land-based wind projects.
- *Applied research to facilitate the integration of wind energy into the electric power grid.* Five initiatives provided funding or support to public or private researchers to conduct applied research related to the integration of wind energy into the electric grid. For instance, DOE's Grid-Scale Rampable Intermittent Dispatchable Storage program funds efforts to develop new technologies that enable widespread use of cost-effective grid-scale energy storage, particularly technologies that mitigate variability in energy generated from renewable sources such as wind and solar. Similarly, the National Science Foundation's Emerging Frontiers in Research and Innovation initiative provides grants for interdisciplinary engineering research with the potential to create a significant impact in meeting national needs. Specifically, in fiscal year 2011, the initiative funded research on compressed air technology for storing excess energy from offshore wind turbines to

alleviate power supply and demand imbalances on the electric grid during the day.[30]

- *Deployment of offshore wind technologies by state, local, and other governmental organizations.* Five initiatives helped address policy and regulatory barriers to deployment of offshore wind through support for state, local, and other governmental organizations. For instance, under Interior's Renewable Energy Program there is a Development and Implementation initiative through which BOEMRE worked to authorize orderly, safe, and environmentally responsible renewable energy development on the outer continental shelf, while complementing ongoing state and local renewable energy efforts. BOEMRE's efforts include assisting states in meeting goals established in RPSs through studies of potential development of particular states' offshore areas, as well as coordination and information exchange with states and regional organizations. Similarly, Commerce's MarineCadastre.gov initiative supports wind energy development and other uses of the outer continental shelf by providing mapping information for project planning and siting, which is intended to help developers identify and avoid potentially conflicting uses before creating development plans. Commerce's efforts under this initiative in fiscal year 2011 included developing maps for a federal-state task force to facilitate decisions on wind energy development in federal waters.

In addition, we identified several types of state initiatives that encourage development of wind and other renewable energy sources and share key characteristics with federal wind-related initiatives. Along with federal agencies, state governments implement initiatives that help energy companies finance deployment of utility-scale land-based wind facilities. These initiatives include state tax incentives such as production and investment tax credits. They also include state grant and loan programs, some of which were federally funded, according to DSIRE data. See figure 4 below for examples of these state and federal initiatives available in fiscal year 2011.

States have also enacted a number of rules, regulations, or policies that encourage deployment of wind and other renewable energy sources, most notably RPSs. Such standards do not provide direct financial support to particular wind projects; however, by requiring or encouraging that a percentage of the electricity consumed in a state be generated from renewable sources, they are designed to create market demand for electricity from sources such as wind. Recent economic studies we reviewed suggest that certain RPSs have increased development of renewable energy.[31] Several financial professionals and agency officials with whom we spoke cited RPSs as strongly influencing wind energy development. They said that by creating demand for wind and other renewable energy, RPSs complement federal initiatives such as the PTC, which reduce the price of this energy. Currently, 37 states have RPSs that include wind.[32] For instance in California—which led the nation in new wind capacity added in 2011—state law requires electric utilities in the state to have 33 percent of their retail sales derived from eligible renewable energy resources, including wind, by 2020. In Texas—which led the nation in total installed wind capacity in 2011—state law requires 5,880 megawatts of total installed renewable energy capacity by 2015, including up to 5,380 megawatts of wind energy capacity.[33]

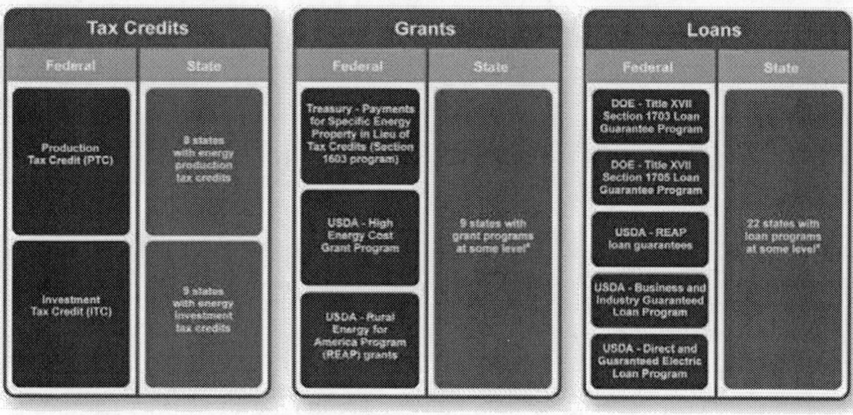

Source: GAO analysis of DSIRE data and agency-provided data.

[a] Though most of these programs are implemented by state governments, several are implemented by one or more utilities or local governments within a state.

Figure 4. Examples of Federal and State Initiatives Available in Fiscal Year 2011 to Provide Financial Support for Deployment of Utility-Scale Land-Based Wind Facilities by Energy Companies.

Overlap among initiatives does not necessarily lead to duplication because initiatives sometimes differ in meaningful ways. For instance, certain of Treasury's tax expenditures that support deployment of utility-scale land-based wind facilities by the energy industry differ in the type of organization eligible for their support. Treasury's Credit for Holding New Clean Renewable Energy Bonds, for example, helps tax-exempt entities such as not-for-profit electric utilities or cooperative electric companies finance capital expenditures for wind and other renewable energy facilities by providing tax credits to holders of bonds issued by those entities. In contrast, the majority of Treasury's other wind-related tax expenditures that support the deployment of utility-scale land-based wind facilities do so by providing taxable organizations with tax credits or other tax incentives for such projects. In addition, certain overlapping initiatives may provide a cumulative benefit for the deployment of wind projects but do not meet our definition of duplication because they do not provide financial support to the same recipient for a single wind project. For instance, Treasury officials noted that the Qualifying Advanced Energy Project Credit and the Section 1603 program could provide support across multiple stages of wind energy deployment if, for example, a manufacturing plant that produces wind turbines receives the credit and a wind facility that uses those turbines receives a Section 1603 grant. However, for purposes of our report, these initiatives do not provide duplicative support because they have different direct recipients.

About Half the Initiatives Reported Formally Coordinating, in Some Cases Employing Key Practices for Enhancing and Sustaining Agency Collaboration

Officials from 43 (52 percent) of the 82 initiatives reported coordinating formally with other federal wind-related initiatives.[34] Coordination was most prevalent among initiatives that reported having wind-specific goals—such as reducing the cost of wind technologies or facilitating the siting, leasing, and construction of new offshore wind projects. Specifically, of the 25 initiatives that reported having such goals, officials from 20 initiatives (80 percent) reported coordinating with other wind-related initiatives. In contrast, of the 57 initiatives that did not report having wind-specific goals, officials from 23 initiatives (40 percent) reported such coordination. Most of the initiatives for which officials reported coordinating—36 of 43—included coordination efforts with wind-related initiatives in other federal departments and independent agencies. For example, officials from several agencies reported coordinating through the Interagency Rapid Response Team for Transmission, which was based on a 2009 joint memorandum of understanding between nine federal agencies. This team aims to improve coordination of federal permitting and reviews of transmission infrastructure projects that will help integrate wind and other renewable energy sources into the electric grid. Officials from

22 initiatives also reported coordinating their wind-related initiatives with state governments.

As we have previously reported, coordination may reduce the risk of unnecessary duplication, and a lack of coordination can waste scarce funds and limit the overall effectiveness of the federal effort.[35] We have also previously reported that while agencies face a range of barriers when they attempt to collaborate with other agencies—such as differing missions and incompatible processes—certain key practices can help agencies enhance and sustain federal collaboration.[36] Several lead agencies implementing wind-related initiatives have formally coordinated overlapping initiatives, in some cases in a manner consistent with key practices, such as in the following examples:

- *Identifying and addressing needs by leveraging resources*: Agency officials reported leveraging resources such as the knowledge and expertise of other agencies in developing and implementing their own wind energy initiatives. For example, DOE has drawn on Treasury's expertise through required consultations regarding the terms and conditions of loan guarantees DOE provides to applicants.[37] In addition, USDA officials consulted with Treasury regarding the potential impact of tax laws on new provisions USDA was drafting for awarding certain grants. For some initiatives, agency officials reported leveraging the financial resources of other agencies to help ensure prudent use of funds, particularly in the case of loan guarantee programs. For example, according to an official for DOE's loan guarantee programs, if a project qualifies for a Treasury Section 1603 grant, it is typical for DOE to require that a portion of the grant proceeds be used to repay the DOE-guaranteed debt.[38] In addition, a USDA official for the Business and Industry Guaranteed Loan Program—which provides guaranteed loans to rural borrowers for projects that improve the economic and environmental climate in rural communities—said that the program provides information to applicants regarding other sources of funding available for wind projects, such as SBA loans or grants, or state-level sources of funding.
- *Developing mechanisms to monitor, evaluate, and report on results*: Agencies also engaged in collaborative efforts to create the means to monitor and evaluate their wind-related initiatives and report on their activities. For example, USDA reported coordinating programs that provide loans, grants, and loan guarantees for projects in rural

communities both internally and with other agencies such as Commerce, DOE, and EPA through USDA's Energy Council Coordinating Committee. Agency officials participating in this council share general information on energy-related programs, which helps support common performance reporting and budgeting processes. In addition, USDA data on its grants, loans, and other awards are available to other agencies and the general public through a web-based mapping tool that shows agencies and potential recipients where USDA is supporting renewable energy projects.

Several Initiatives That Financed Deployment of Wind Facilities Have Provided Some Duplicative Financial Support

We identified examples of utility-scale land-based wind projects that received duplicative support—financial support from multiple initiatives to the same recipient for a single project—from initiatives supporting deployment of wind facilities. Specifically, we identified 10 initiatives that have provided or could provide duplicative support, as follows:[39]

- Seven initiatives provided some duplicative support for wind projects, including three tax expenditures and a grant program implemented by Treasury, a loan guarantee program implemented by DOE, and two programs that provide grants, loans, and loan guarantees implemented by USDA. Although all seven of these initiatives cannot be combined to support the same project, each of them has been combined with one or more of the others, with some limitations, to support a single project.
- Three other initiatives did not actually fund any wind projects in fiscal year 2011 but could provide duplicative support for wind projects going forward, based on the types of projects eligible for their support. These three initiatives were DOE's Section 1703 Loan Guarantee Program (Section 1703 program) and USDA's Business and Industry Guaranteed Loan Program and High Energy Cost Grant Program.

Some of these 10 initiatives have recently expired, such as Treasury's Section 1603 program and DOE's Section 1705 Loan Guarantee Program (Section 1705 program), and several others are scheduled to expire for wind projects at the end of 2013, such as Treasury's tax credits.[40] However, the types of mechanisms these initiatives employ—tax expenditures, grants, and loan guarantees—are employed by other of the initiatives that are not expiring and may be considered by policymakers as a means for supporting wind energy through future initiatives. In addition, duplication of financial support among these initiatives may not be limited to wind projects because all of these initiatives supported a range of renewable energy projects. Of the 10 initiatives, Treasury's 4 initiatives accounted for over 95 percent of the total federal financial support for wind in fiscal year 2011. See table 4 for brief descriptions of these initiatives. (For more detailed descriptions of these initiatives, including information on their expiration dates, see app. II.)

Table 4. Ten Initiatives That Have Provided or Could Provide Duplicative Financial Support for Deployment of Wind Facilities

Agency	Initiative	Description
DOE	Title XVII Section 1703 Loan Guarantee Program (Section 1703 program)[a] Title XVII Section 1705 Loan Guarantee Program (Section 1705 program)	According to DOE, the Section 1703 program provides loan guarantees to support innovative clean energy technologies that are typically unable to obtain conventional private financing due to high technology risks. The law requires that the technologies avoid, reduce, or sequester air pollutants or emissions of greenhouse gases. The Section 1705 program was a temporary program providing loan guarantees for both innovative and commercial technology energy projects that employ wind and other renewable energy systems, electric power transmission systems, or leading-edge biofuels that meet certain criteria. All Section 1705 projects were required to begin construction no later than September 30, 2011. Following the expiration of the Section 1705 program, Congress appropriated $170 million to pay the credit subsidy costs for Section 1703 projects that use renewable energy or efficient end-use energy technologies. The law provides that this funding is also available to such projects that applied under the Section 1705 program prior to February 24, 2011.

Table 4. (Continued)

Agency	Initiative	Description
Treasury	Energy Production Credit (PTC) Energy Investment Credit (ITC) Payments for Specific Energy Property in Lieu of Tax Credits (Section 1603 program)	The PTC provides an income tax credit based on the amount of energy produced at qualified facilities, including wind facilities. *As an alternative to the PTC*, the ITC provides an income tax credit of 30 percent of either the cost or fair market value of new equipment that produces electricity from wind and other renewable energy sources. The payments under the Section 1603 program, *which can be taken in lieu of the PTC or ITC*, provide cash grants worth 30 percent of a wind project's cost or fair market value.
	Accelerated Depreciation Recovery Periods for Specific Energy Property (accelerated depreciation)[b]	Allows wind energy technologies to be treated as 5-year property—that is, property whose costs are recovered through depreciation deductions from businesses' taxableincome over 5 years. The Joint Committee on Taxation generally classifies as tax expenditures cost recovery allowances that are more favorable than those provided under the alternative depreciation system (Internal Revenue Code Section 168(g)), which provides for straight-line recovery over tax lives that are longer than those permitted under the accelerated system. Accelerated depreciation, in effect, reduces the cost of acquiring wind and other properties by allowing businesses to deduct larger amounts from their taxable income sooner than they would be able to do understraight-line depreciation. Reducing tax liability earlier provides a benefit to the taxpayer because of the time value of money—having a lower tax payment today is worth more to the taxpayer than having the lower payment in the future.
USDA	Business and Industry Guaranteed Loan Program[a]	Provides guaranteed loans to borrowers in rural areas for a range of eligible projects that improve the economic and environmental climate in rural communities. Eligible activities include the development and construction of renewable energy systems.

Agency	Initiative	Description
	Direct and Guaranteed Electric Loan Program	Provides loans and loan guarantees for a range of eligible projects that establish and improve electric service in rural areas, including renewable energy systems.
	High Energy Cost Grant Program[a]	Provides grants for energy generation, transmission, and distribution facilities servingrural communities with annual average home energy costs that exceed 275 percent of the national average. Eligible projects include on-grid and off-grid renewable energy systems.
	Rural Energy for America Program (REAP)	Provides funding for grants and guaranteed loans to farmers, ranchers, and small businesses in rural areas to assist with purchasing and installing renewable energy systems, such as wind projects.

Source: GAO analysis of agency-provided data.

Note: Each of these 10 initiatives has been or could be used together with 1 or more other initiatives in this table; however, recipients can receive support from only 1 of the following 3 initiatives for a specific project: the PTC, ITC, or Section 1603 program.

[a] The Section 1703 program has not funded any wind or other projects to date, though it has provided conditional commitments to guarantee over $10 billion in loans for nuclear energy projects. However, recipients of Section 1703 loan guarantees are not restricted from receiving support from tax initiatives, such as Treasury's tax credits, and may receive support from such initiatives in the future. In addition, neither the Business and Industry Guaranteed Loan Program nor the High Energy Cost Grant Program funded any wind projects in fiscal year 2011. However, both initiatives specify wind projects as eligible for funding, and USDA officials said that neither initiative restricts their recipients from receiving support under the DOE or Treasury initiatives listed here.

[b] Depreciation—a normal business expense under an income tax system—is an annual deduction from income that allows taxpayers to recover the cost or other basis of certain property used in a business or other income-producing activity over the useful life of the property. In addition to the existing 5-year accelerated depreciation allowed for wind and other properties, 2008 legislation and subsequent laws have temporarily granted a 50 percent first-year bonus depreciation for properties placed in service before January 1, 2014. This allows businesses to deduct 50 percent of the depreciable basis of a broad set of tangible properties, including wind and other renewable energy facilities, from their taxable income in the first year after they are acquired. Furthermore, the Tax Relief, Unemployment Insurance Reauthorization, and Job Creation Act of 2010 allowed businesses to deduct 100 percent of the depreciable basis of eligible wind and other facilities from their taxable income after September 8, 2010 and before January 1, 2012. The 50 percent bonus depreciation allowed under the 2008 act narrowed any tax differences between eligible assets, and the 100 percent bonus depreciation introduced in 2010 eliminated those differences altogether under the provision for allowing a full write-off of asset acquisition costs.

According to interviews with agency officials and financial professionals and information from agency websites, wind project developers have used various combinations of these initiatives to help finance specific wind projects. For instance, in many cases, developers combined the support of more than one Treasury initiative and, in some cases, they received additional support from smaller DOE or USDA grant or loan guarantee programs. Among Treasury initiatives, although the PTC, Energy Investment Credit (also known as the Energy Investment Tax Credit, or ITC), and Section 1603 program cannot be combined for a specific project, they all support wind projects for which developers also typically claim Accelerated Depreciation Recovery Periods for Specific Energy Property (accelerated depreciation), according to financial professionals and a Treasury official. In addition, DOE's Section 1705 program has provided loan guarantees for four utility-scale wind generation projects, all of which have received grants under Treasury's Section 1603 program, and all of which are eligible to claim accelerated depreciation.[41] Similarly, USDA's grant and loan programs have supported projects that also received support under other initiatives. For example, USDA's Direct and Guaranteed Electric Loan Program provided a loan guarantee for a $204 million loan for a wind project that was also awarded an $88 million grant under the Section 1603 program. In addition, USDA's Rural Energy for America Program (REAP) provides grants and loan guarantees for renewable energy projects. A 2006 report by DOE's Lawrence Berkeley National Laboratory (LBNL) found that nearly all wind projects with a capacity of over 100 kilowatts that received REAP grants from 2003 through 2005 also intended to claim tax credits under the PTC.[42]

Although these initiatives have, in some cases, provided duplicative support, their support may address different needs of wind project developers or the communities they serve, according to agency officials and financial professionals with whom we spoke, and analyses by DOE's national laboratories.[43] For example, unlike the PTC and ITC, the Section 1603 program allows wind project developers to directly claim a cash grant regardless of their tax liability, thus avoiding the potential need to partner with financial institutions or other investors who provide tax equity.[44] The Section 1603 program was created in part to address a perceived lack of tax equity following the recent financial crisis, according to Treasury guidance and financial professionals. Furthermore, by providing a cash grant, the program allows developers to receive the full amount of the government subsidy rather than sharing this subsidy with tax equity investors.[45] DOE's Section 1705 program, meanwhile, provided financing in many cases for innovative projects that were seen as too risky to obtain affordable financing from private lenders, according to DOE officials who administered the program. In addition, as with the Section 1703 program, Section 1705 loan guarantees can address projects' needs for construction and long-term debt financing, while grants under the Section 1603 program and support from Treasury's tax expenditures are available only when the related project has been constructed and is operational. Therefore, the loan guarantees helped support many projects that might not otherwise have reached the development stage—such as being placed in service or beginning to generate electricity—required to receive tax credits or Section 1603 grants. In addition, USDA's grant, loan, and loan guarantee initiatives are designed to encourage projects that serve the needs of rural communities, including by providing reliable, affordable electricity, and more generally stimulating rural economic development.

Moreover, although these initiatives can be used together in various combinations to help finance the same wind project, several include provisions—often referred to as antidouble-dipping provisions—that limit the amount of financial support provided to a wind project when combined with another initiative. For instance, the PTC includes a provision requiring that the amount of the credit be reduced for federal or state grants, tax-exempt bond financing, subsidized energy financing, or other federal tax credits received for use in connection with the project.[46] LBNL's financial modeling of wind projects for its 2006 report suggests that large wind projects receiving REAP grants and claiming the PTC would have seen the value of the PTC reduced by from 11 to 46 percent of the grant's face value, depending on the project's capital cost and capacity factor.[47] Similarly, antidouble-dipping provisions reduce the value of the ITC and Section 1603 grants—through reductions to the portion of project costs on which they are calculated—for projects that also receive government grants that are not taxed as income,[48] including, in some cases, REAP or other federal and state grants for wind projects.[49] Grants not taxed as income also reduce a project's depreciable basis, or the dollar amount that can be depreciated for tax purposes.[50] In addition to limitations on combining Treasury's tax expenditures with other sources of financial support, officials from USDA and DOE told us that their agencies consider some other sources of federal support a wind project has received or will receive in determining whether or how much to award under their grant, loan, and loan guarantee programs. The officials for some of these programs said that they limit the value of support they provide, while officials from other programs, by law, must deny support altogether when applicants are receiving funding from other federal sources. For instance, the appropriations laws applicable to the

Section 1703 program prohibit the issuance of loan guarantees for projects that are expected to receive certain other sources of federal support. Such sources of support include grants from certain USDA initiatives and federal "off-take arrangements," whereby federal agencies agree to purchase power from the projects.[51] Similarly, USDA officials said that, under REAP, the total amount of grants provided for projects from REAP and other federal sources generally cannot exceed 25 percent of project costs. However, this limit does not apply to grants provided under Treasury's Section 1603 program, according to USDA officials.

In addition to these 10 federal initiatives, the state tax credits, grant programs, and loan programs previously discussed can be used, in some cases, to provide financial support for deployment of a wind project in combination with one or more federal initiatives. For instance, it is possible for a single wind project to receive federal support from a Section 1603 grant, accelerated depreciation, and a DOE loan guarantee, along with state support from tax incentives and indirect subsidies due to a state RPS. Furthermore, DSIRE staff and a financial professional with whom we spoke said that states may often structure their initiatives so that recipients can fully leverage sources of federal support, such as by designing the initiatives to avoid triggering federal antidouble-dipping provisions. In addition, under Treasury's published guidance on the PTC provision reducing the amount of the credit for certain other sources of federal or state support, state or local tax credits do not trigger a reduction in the value of the PTC.[52]

Even with antidouble-dipping provisions and other limitations on combining financial support from multiple initiatives for the same project, federal initiatives have provided cumulative financial support worth about half of project costs for many wind projects according to financial professionals. For instance, financial professionals we spoke with estimated that the PTC and accelerated depreciation together provide nearly half of the capital costs required for a typical wind farm.[53] Of this amount, 30 percent or more of the total capital costs is due to the PTC, according to financial professionals' estimates.[54] For projects receiving support from other federal grant or loan initiatives in addition to the PTC and accelerated depreciation, the value of federal financial support would comprise a larger portion of project costs. Also, as noted earlier, wind projects may receive financial support from state initiatives. For instance, according to a briefing memorandum from White House staff, the total estimated federal and state financial support for a large wind project in Oregon—including a Section 1705 loan guarantee, a Section 1603 grant, accelerated depreciation at the federal and state level, state tax credits, and an estimated premium paid for power due to a state RPS—are worth 65 percent or more of the project's capital costs. In another example, estimates developed by management consultants for the energy industry and other clients suggest that federal financial support for a New Hampshire wind farm—including a Section 1603 grant, a Section 1705 loan guarantee, and accelerated depreciation—in combination with financial support from state initiatives is worth over half of the project's capital costs.[55]

Agencies Support Projects on the Basis of Initiatives' Goals or Eligibility Criteria, but the Extent to Which Agencies Assess Applicant Need Is Unclear

Agencies implementing the 10 initiatives that have provided or could provide duplicative support allocate support to projects on the basis of the initiatives' goals or eligibility criteria, but the extent to which agencies assess applicant need for the support is unclear because we found they do not document assessments.[56] DOE and USDA—which have discretion, to the extent allowed by their statutory authority, over the projects they support through 6 of the 10 initiatives—allocate support to projects based on the projects' ability to meet initiative goals such as reducing emissions or benefitting rural communities, as well as other criteria such as financial and technological feasibility. Treasury, meanwhile, provides support to projects through the remaining 4 initiatives based on the eligibility criteria in the tax code. DOE and USDA consider applicant need for the financial support of some initiatives, according to officials. However, we found that neither agency documents assessments of applicant need for any of their initiatives; therefore, the extent to which they use such assessments to determine how much support to provide is unclear. Treasury does not generally have discretion in allocating support to projects and, as such, does not assess need for the support of its initiatives. While the support of these initiatives may be necessary, in many cases, for wind projects to be built, because the agencies do not document assessments of need, it is unclear, in some cases, whether the entire amount of federal support provided was necessary to build wind projects. In the event that some wind projects receive more federal funding than is required to induce them to be built, this additional funding could potentially be used to induce additional projects to be built or simply withheld, thereby reducing federal expenditures.

DOE and USDA Allocate Support to Projects Based on Initiative Goals and Other Criteria while Treasury Supports Projects Based on Eligibility under the Tax Code

Through their six initiatives, DOE and USDA allocate support to projects based on projects' ability to meet initiative goals, along with other criteria such as financial and technological feasibility. For instance, DOE's loan guarantee solicitation for its Section 1703 program set forth initial screening criteria for projects including that they employ new or significantly improved technology compared to commercially available technologies, and that they be ready to proceed through the loan approval process (i.e. equity has been committed to the project, construction and other contracts have been negotiated, and permits have been secured). For evaluating and scoring projects that meet the initial screening criteria, DOE's solicitation also set forth two equally weighted criteria related to the program's goals: a project's expected reduction or avoidance of greenhouse gas emissions relative to its cost, and a project's support for clean energy jobs and manufacturing. In line with program goals, USDA allocates the support of its four initiatives to projects based on their expected benefits for rural and other eligible communities, along with other factors such as technological feasibility and expected performance. For instance, under its High Energy Cost Grant Program, USDA evaluates which projects to support based on their abilities to address community needs such as those related to economic hardships, their technological design and feasibility, their expected performance measures including the amount of renewable energy they will produce, and other factors. Similarly, USDA evaluates applications for REAP grants or loan guarantees based on factors such as projects' support for small agricultural producers or businesses, expected energy production, and technical merit. As with DOE's Section 1703 program, USDA's loan guarantee programs also allocate support to projects based on their ability to repay their debt. Unlike DOE and USDA, Treasury generally does not have any discretion regarding which projects receive the support of its initiatives.[57] Taxpayers who are eligible for support under the Internal Revenue Code are generally entitled to that support.

It Is Unclear to What Extent Agencies Assess Applicant Need for the Financial Support of Their Initiatives Because We Found Such Assessments Are Not Documented

According to agency officials and program guidance, DOE and USDA consider applicant need for the financial support of some of their initiatives.[58] For instance, the solicitation for loan guarantee applications under DOE's Section 1703 program states that DOE will view unfavorably applications for projects that could be fully financed on a long-term basis by commercial banks or others without a federal loan guarantee. DOE officials told us they require applicants to provide a letter stating whether their projects can be financed without a federal loan guarantee, although this self-certification by applicants does not require they document any efforts to obtain private financing.[59] In addition to these letters, DOE officials said their conversations with wind project developers, along with their broader understanding of the lending community and project risks, allow them to determine whether projects would likely be able to obtain private financing without a loan guarantee. USDA also considers applicants' need for support from some of its initiatives according to agency officials. For example, application guidance for USDA's High Energy Cost Grant Program states that the program assesses project information including other sources of funding expected for the project to determine its financial viability, the level of community support for the project, and the community's need for funds. Similarly, officials from the Direct and Guaranteed Electric Loan Program said that prior to loan approval they assess projects' financial information to determine their financial feasibility and to avoid lending more than is necessary for project completion. However, unlike DOE's Section 1703 program, we did not identify program documents for USDA's initiatives—such as guidance for applicants or criteria for evaluating projects—stating that applicant need is a factor in allocating support. Treasury generally does not have any discretion in allocating support to projects and, as such, does not assess applicant need for the support of its initiatives.

Even with these DOE and USDA efforts, it is unclear to what extent DOE and USDA assess applicant need for the financial support of their initiatives because we found they do not document such assessments. The federal standards for internal control include control activities—such as documentation of significant transactions—which are essential for proper stewardship and accountability for government resources.[60] Because, as we found, the agencies do not document these assessments, it is unclear to outside parties how they considered the financial need of applicants when determining what amount of support to provide. Moreover, it is unclear if the incremental support some initiatives provided was always necessary to build projects. In the event that some wind projects receive more federal funding than is required to induce them to be built, this additional funding could potentially be used to induce additional projects to be built or simply withheld, thereby reducing federal expenditures. The following are examples where it was unclear whether initiatives' incremental support was needed for projects to be built[61]:

- According to the White House briefing memorandum noted above, because of the tax subsidies and other federal and state support for the Oregon wind project, the return on the private equity invested in the project was estimated to be relatively high—around 30 percent.[62]

The memorandum further stated that this estimated return suggests the project would "likely move without the [Section 1705] loan guarantee," and "the alternative of private financing would not make the project non-viable." It is unclear from our review whether the loan guarantee was needed for the project to be built because we found DOE made no documented assessment of need. In addition, a separate analysis of the same wind project by DOE suggested that officials concluded, given the amount of the project's debt, it would have sufficient cash flow to repay its guaranteed loan without the incremental support of a Section 1603 grant, which it later received. Specifically, DOE approved a loan guarantee for the project in part based on its credit analysis, which was made under the assumption that the project would not need to make use of a Section 1603 grant to repay debt, and that neither DOE nor lenders for the project would have any claim on the grant. However, it is unclear whether the Section 1603 grant was needed for the project to be built because we found no documented assessment of need was made.[63] Though the analyses from the White House memorandum and DOE question the project's need for the combined support of the Section 1705 loan guarantee and Section 1603 grant, neither analysis questioned whether the project would have been built without either source of support.

- In another example, a developer of a wind project in Maine provided documentation in 2011 that it had sufficient funds to complete construction of the project without any additional source of capital, though it subsequently received a Section 1603 grant and a Section 1705 loan guarantee and was eligible for accelerated depreciation.

> Specifically, the developer provided this information to document its financial capacity in support of its permit application to the Maine Department of Environmental Protection, which later approved the permit for the project. However, because we found no documented assessments of need were made for the federal support this project received, it is unclear whether it could have been built with less support.

Nonetheless, the incremental support agencies' initiatives provide may be necessary for wind projects to be built, according to agency officials and financial professionals. For instance, concerning DOE's initiatives, its loan guarantees allow developers to leverage federal resources to attract sources of equity and debt that would otherwise not be invested, according to DOE. Officials from the loan guarantee programs said that, without loan guarantees, wind project developers can have difficulty obtaining private loans due to the relatively long term of the fixed rate loans they use to finance their projects. Title XVII of the Energy Policy Act of 2005 allows DOE to provide guarantees for loans with terms of up to 30 years. According to DOE officials, there are constraints on the supply of private financing for large projects, and private lenders may consider such long-term loans to have greater risks and may be less likely to lend to such projects in the amounts required to fully finance the transactions.

In addition, in commenting on a draft of this report, DOE officials said that long-term financing is necessary in order for debt payments to align with projects' proceeds from agreements to sell their power over the long term, and is also necessary to avoid risks associated with changing interest rates and other risks that can arise from using shorter-term financing. For instance, they said that the Maine wind project developer's filings with the state did not address any long-term financing needs for the project beyond its construction phase.

With respect to USDA's initiatives, USDA officials for some initiatives told us that their incremental support may be necessary for wind projects to be built, and for the projects to fully benefit rural communities. For instance, officials noted that, although well-qualified projects can generally find the financing they need in the private market, the cost of private financing would be higher than the cost of financing available through USDA's loan and loan guarantee programs, which would likely impact electric utility rates for rural ratepayers. They also said that projects receiving support from the High Energy Cost Grant Program may not be built without its support, as it tends to serve isolated communities where available funding for such projects may be limited.

Regarding the PTC and other Treasury initiatives, several financial professionals with whom we spoke said that the initiatives provide financial support for many projects that would likely not be built otherwise. For instance, they said that the PTC is necessary in order for many wind projects to be financially viable.

Furthermore, although it was extended in 2013, prior to this extension, the financial professionals said the PTC's scheduled expiration at the end of 2012 had caused developers and investors to suspend plans for future construction of or investment in wind projects.[64] This expected slowdown in deployment of new wind projects is in line with historical evidence of prior PTC expirations being followed by decreases in new wind energy capacity additions. Treasury's Section 1603 program has also been shown to support wind projects that would otherwise likely not have been built. According to an LBNL analysis, the program supported projects that would likely not have been built using the PTC if the grant were not available—projects that added as much as 2,400 megawatts of wind energy capacity in 2009.[65]

More broadly, according to financial professionals, wind project developers and investors evaluate the returns they could make on a range of potential projects. If the expected returns for wind projects are lower due to a decrease in federal support, developers and investors are more likely to pursue other types of projects—including solar or other renewable energy projects, as well as nonenergy projects—that benefit from federal subsidies and could provide higher returns.[66]

CONCLUSION

Faced with concerns about the nation's reliance on imported oil, as well as fossil fuels' contribution to global climate change, among other things, federal policymakers have increased the federal focus on and support for development of renewable energy sources, especially wind energy.

At the same time, states have created demand for energy from renewable sources through initiatives such as RPSs, supplementing support provided by federal agencies.

In fiscal year 2011, wind-related initiatives implemented by federal agencies were fragmented and, in many cases, overlapping. Further, we identified 10 initiatives that have provided or could provide duplicative support to deploy wind facilities. Though some of the 10 initiatives have recently expired or are scheduled to expire, other initiatives employing similar mechanisms such as tax expenditures, grants, and loan guarantees remain in place, and similar initiatives may be considered in the future as a means for supporting wind and other renewable energy sources.

In the current fiscally constrained environment, it is especially important to allocate scarce resources where they can be most effective. In this context, it is important that agencies with discretion in implementing initiatives that have provided or could provide duplicative support—DOE and USDA—ensure that they allocate support through these initiatives to projects that would not be built otherwise.

However, these agencies do not make documented assessments of whether or how much of their initiatives' financial support is needed for projects to be built and, as a result, it is unclear to what extent they assess need in order to determine what amount of support to provide. Moreover, it is unclear whether the incremental support some initiatives provided was always necessary for wind projects to be built.

Recommendation for Executive Action

To support federal agencies' efforts to effectively allocate resources among wind projects, we recommend that the Secretaries of Energy and Agriculture, to the extent possible within their statutory authority, formally assess and document whether the incremental financial support of their initiatives is needed in order for applicants' projects to be built and take this information into account in determining whether, or how much, support to provide.

Such assessments could include, for example, information on the investors' and developers' projected rates of return on these projects, or documentation of applicants' inability to secure private financing for projects. In addition, such assessments should consider the financial support available or provided to projects from other federal sources including tax expenditures and, to the extent practical, from state sources. In the event agencies lack discretion to consider this information in determining what financial support to provide, they may want to report this limitation to Congress.

Agency Comments and Our Evaluation

We provided a draft of this report to the Secretaries of Energy, Agriculture, and the Treasury for review and comment. DOE provided written comments, in which it agreed with our recommendation; these comments are summarized below. USDA's Rural Development provided comments by e-mail on February 11, 2013, stating that USDA generally concurred with the information in our report related to its initiatives. In addition, DOE, USDA, and Treasury provided technical and clarifying comments, which we incorporated as appropriate.

DOE stated in its written comments that it will now formally document its evaluation of applicants' assertions regarding their inability to finance their projects without a federal loan guarantee, and it will clarify how it considers the financial need of applicants when determining what amount of support to provide. With regard to financing wind projects, DOE noted that Section 1603 grants do not provide capital for developers to use to construct projects, but rather the proceeds from the grants are only available when the related project construction is complete and the project is operational. In contrast, DOE noted that its loan guarantees provide construction and long-term debt financing. As we note in the report, these initiatives may address different needs of wind project developers, including the need for project financing prior to reaching the development stage required to receive tax credits or grants under the Section 1603 program. To emphasize DOE's point, however, we added language to the report to make it clear that grants do not provide project sponsors with capital to construct their projects.

Frank Rusco
Director, Natural Resources and Environment

Appendix I: Objectives, Scope, and Methodology

Our objectives were to (1) identify wind-related initiatives implemented by federal agencies in fiscal year 2011 and their key characteristics; (2) assess the extent of fragmentation, overlap, and duplication, if any, among these initiatives, and the extent to which they were coordinated; and (3) examine how agencies allocate support to projects through their initiatives and the extent to which they assess applicant need for support.

To inform our objectives, we reviewed our February 2012 report that identified federal agencies' renewable energy initiatives and examined the federal roles the agencies' initiatives supported in fiscal year 2010.[1] That report identified nearly 700 initiatives that were implemented across the federal government, of which 296 initiatives supported wind energy.[2] For purposes of this report, we generally only included those wind-related initiatives categorized under the research and development or commercialization and deployment federal roles in our February 2012 report. However, we included some initiatives categorized under the regulation, permitting, and compliance federal role if they had a clear focus on deployment of wind energy, such as through streamlining regulatory processes or fast-tracking permitting or other processes for wind projects.[3] From this list of initiatives, we excluded those at certain agencies—such as the Departments of Defense, Homeland Security, and State—whose initiatives generally focused on development of wind energy and other technologies for use in a military, border security, or international aid setting, rather than for use in the domestic commercial energy market.

For the remaining agencies and initiatives, we developed an initial questionnaire to collect information from officials regarding whether the fiscal year 2010 initiatives were still active and whether wind energy still received or was eligible for support under the initiatives in fiscal year 2011.[4] We also asked officials to identify any additional initiatives that were active and for which wind energy was eligible for support in fiscal year 2011. If officials wanted to remove an initiative from our list, we asked for additional information to support the removal. Using the responses from this questionnaire, we identified 82 wind-related initiatives at nine agencies.

To identify and describe the key characteristics of wind-related initiatives implemented by federal agencies, we developed a second questionnaire to collect information from officials responsible for the 82 initiatives. The questionnaire was prepopulated with information that was obtained from the agencies for GAO's report on fiscal year 2010 renewable energy initiatives, including the initiative name, description, recipient type, and expiration or sunset date. We asked officials to confirm or modify this information as appropriate for fiscal year 2011. We requested additional information on the initiatives including their obligations, or revenue losses from tax expenditures[5] for activities specifically related to wind;[6] year in which they first supported wind energy; type of wind issues and technology advancement activities supported; initiative-wide and wind-specific goals; and efforts to coordinate with other wind-related initiatives. For a copy of our questionnaire, see appendix IV. We conducted pretests with officials from 12 initiatives across three agencies to ensure that respondents interpreted our questions in the way we intended (e.g., the questions were clear and unambiguous and terminology was used correctly), that the questionnaire was comprehensive and unbiased, and that respondents had the necessary information and ability to respond to the questions. An independent GAO reviewer also reviewed a draft of the questionnaire prior to its administration. On the basis of feedback from these pretests and independent review, we revised the questionnaire in order to improve its clarity. After completing the pretests, we sent the finalized questionnaires to the appropriate agency liaisons, who in turn sent the questionnaires to the appropriate officials. We received questionnaire responses for each of the 82 initiatives, resulting in a response rate of 100 percent. After reviewing the responses, we conducted follow-up e-mail exchanges or telephone discussions with agency officials when responses were unclear or conflicting. When necessary, we used the clarifying information provided by agency officials to update answers to questions to improve the accuracy and completeness of the data. To assess the reliability of obligations data, our questionnaire included questions on the data systems used to generate that data and any methodologies agencies used to develop estimates of obligations for their initiatives. In addition, to assess the reliability of data on tax subsidies provided by wind-related tax expenditures, we interviewed officials from the Department of the Treasury regarding the how the data were developed, and compared the data between the two publicly available sources from the Joint Committee on Taxation and the Office of Management and Budget. We determined that the obligations and tax subsidy data used in this report were of sufficient quality for our purposes. Because this effort was not a sample survey, it has no sampling errors. However, the practical difficulties of conducting any survey may introduce errors, commonly referred to as nonsampling errors. For example, difficulties in interpreting a particular question, sources of information available to respondents, or entering data into a database or analyzing them can introduce unwanted variability into the survey results. However, we took steps to minimize such nonsampling errors in developing the questionnaire—including using a social science survey specialist to help design and pretest the questionnaire. We also minimized the nonsampling errors when analyzing the data, including using a computer program for analysis, and using an independent analyst to review the computer program. Finally, we verified the accuracy of a small sample of keypunched records by comparing them with their corresponding questionnaires, and we corrected the errors we found. Less than 0.5 percent of the data items we checked had random keypunch errors that would not have been corrected during data processing.

To assess the extent of fragmentation, overlap, and duplication of wind-related initiatives, we first defined these terms based on definitions established in our prior reports. Specifically, for purposes of this report, fragmentation, overlap, and duplication, were defined as follows:

- Fragmentation occurs when more than one federal agency, or more than one organization within an agency, is involved in the same broad area of national need.
- Overlap occurs when multiple initiatives support similar wind issues, similar technology advancement activities, and similar recipients, as well as having similar goals.
- Duplication occurs when multiple initiatives provide financial support to the same recipient for a single wind project.[7] Duplication as we have defined it may be necessary in some cases for specific wind projects to be built. However, in other cases, duplication may result in ineffective use of federal financial support—that is, it may result in some amount of support being provided for specific wind projects that is not needed for them to be built.

To determine the extent of fragmentation, we used agencies' questionnaire responses to confirm the number of federal agencies that supported wind-related initiatives. To determine the extent of overlap, we first analyzed the questionnaire responses to categorize initiatives' recipient types into four categories, as follows[8]:

- *Energy providers, developers, and manufacturers*. This category includes organizations in the energy industry that provide electricity produced by wind energy, develop wind energy generation projects, or manufacture equipment associated with wind turbines or other wind-related technologies.
- *Public and private researchers*. This category includes researchers employed by or associated with federal, state, or other governmental entities (such as national laboratories), academic institutions, nonprofit organizations, or private companies.
- *State, local, tribal, and other governmental organizations*. This category includes nonfederal governmental organizations, such as

state and local governments and quasi-governmental entities, and federally recognized American Indian tribes.

- *Individuals*. This category includes members of the general public who produce, develop, or use wind energy, and who receive support independently of their affiliation with a private, governmental or other organization.

We then analyzed information on the initiatives' descriptions and goals provided in the questionnaire responses and categorized initiatives into all applicable categories that we developed for types of goals. These categories included initiatives that facilitated the assessment of wind resources; initiatives that fostered technological improvements or cost-reduction in wind technologies; initiatives that financed the construction or use of wind facilities; and initiatives that addressed policy and regulatory barriers to wind energy development. Once these categories were defined, two staff independently read through each initiative's description and goals and identified all categories that likely applied to the initiative. They then discussed the categorizations about which they disagreed and came to agreement about whether or not the category applied to the initiative. Using agency-provided data on wind issues and technology advancement activities supported, and our categorizations of the initiatives' recipients and types of goals, we identified overlapping initiatives as those sharing at least one common wind issue, technology advancement activity, recipient type, and type of goal.

To identify duplication of federal support, we focused our review on those initiatives with the largest estimated obligations or revenue losses in 2011 for activities specifically related to wind.

Specifically, we first focused our analysis on the five lead agencies, which implemented 89 percent of initiatives comprising 99.9 percent of estimated obligations and all estimated revenue losses in 2011—the Departments of Energy (DOE), the Interior, Agriculture (USDA), Commerce, and the Treasury.

Second, we focused our analysis on initiatives that included support for deployment, which were responsible for 99 percent of obligations and all estimated revenue losses in 2011.

Third, because of the relatively large number of and variety in the initiatives that supported deployment, we further focused our analysis on those deployment initiatives that provided financial support for construction or operation of wind facilities. Fourth, we narrowed our focus to initiatives that included a focus on utility-scale land-based wind—the most commonly supported wind issue—and fifth, we narrowed our focus to initiatives with recipients that included energy providers, developers, or manufacturers—the most commonly support recipient type. Applying all of these criteria resulted in a list of 15 initiatives, which represented 96 percent of estimated obligations and all revenue losses, according to best available estimates. From this list of 15 initiatives, we reviewed agencies' questionnaire responses, agency documents, and laws and regulations related to the initiatives, and spoke with agency officials and outside experts about them. Based on this review, we determined that there was only a small potential that duplicative support was provided by the four Treasury initiatives because eligibility for their support was explicitly limited to tax-exempt entities, which were generally not supported by other initiatives such as Treasury's other tax expenditures. In addition, on the basis of our review of documents and discussions with agency officials and others, we determined that there was only a small potential for duplication of another initiative on our list—Treasury's Qualifying Advanced Energy Project Credit—because its eligibility criteria limit its support to manufacturing facilities, rather than the energy generation facilities that are generally supported by the other initiatives we identified that have provided or could provide duplicative support. In addition, all available credits under the initiative were allocated by the end of 2010.

For initiatives we identified that have provided or could provide duplicative support, we collected information from agency websites on financial support provided for projects, and we interviewed agency officials and reviewed program guidance and regulations for information on how agencies allocate support to projects through the initiatives, and efforts by the agencies to assess applicant need for the support of their initiatives. We also reviewed studies of the initiatives by DOE's national laboratories, the Congressional Research Service, and other experts. In addition, we interviewed six financial professionals from several of the major financial institutions and legal firms active in wind energy project financing in recent years regarding the support for wind projects provided by the initiatives. We identified these individuals based on their presentations at the annual national wind industry conference held by the American Wind Energy Association and through reviews of industry reports, newsletters, and other publications. To obtain additional information about the types of support available to wind project developers from state governments, we collected and analyzed data from the Database of State Incentives for Renewables and Efficiency (DSIRE), a comprehensive source of information on state incentives and policies that promote renewable energy and energy efficiency, which is funded by DOE. We interviewed researchers who developed and maintain DSIRE regarding their methodology for collecting and summarizing information on state incentives and policies and their processes for ensuring the data are accurate and up-to-date, and we determined the data were sufficiently reliable for our purposes. We also interviewed agency officials and financial professionals for additional information on state initiatives.

We conducted this performance audit from February 2012 to March 2013 in accordance with generally accepted government auditing standards. Those standards require that we plan and perform the audit to obtain sufficient, appropriate evidence to provide a reasonable basis for our findings and conclusions based on our audit objectives. We believe that the evidence obtained provides a reasonable basis for our findings and conclusions based on our audit objectives.

APPENDIX II: FEDERAL WIND-RELATED INITIATIVES

Tables 5 through 13 below provide descriptions, by agency, of the 82 federal wind-related initiatives we identified. The tables also provide information reported by agencies on initiatives that will or have expired, in full or in part, due to an expiration of legislative authority, depletion of available appropriations, or some other expiration under the law as of fall of 2012.

Table 5. U.S. Department of Agriculture (USDA) Wind-Related Initiatives

Initiative name and implementing subagency	Description	Expiration information
Agricultural Research Service		
Bioenergy National Program	The mission of this Agricultural Research Service research program is to develop technologies for the sustainable commercial production of biofuels by the agricultural sector in ways that enhance natural resources without disrupting existing food, feed, and fiber markets. Agricultural Research Service efforts under this initiative include a research project involving wind energy.	Wind component will be phased out completely by fall of 2014.

Initiative name and implementing subagency	Description	Expiration information
U.S. Forest Service		
Landownership Management Program	Under this program, the U.S. Forest Service manages over 193 million acres of forests and grasslands and provides approximately 70,000 land use authorizations annually for a wide variety of purposes, including rights-of-way for roads, pipelines, communication and navigation sites, and electric transmission and distribution facilities. To help meet USDA's strategic goals of assisting rural communities and federal energy goals, the U.S. Forest Service is working to diversify the energy projects for which it issues land use authorizations to include projects that produce energy from renewable sources such as wind, solar, and hydroelectric power.	None
National Institute of Food and Agriculture		
Small Business Innovation Research Program: Rural Development Topic Area	As part of USDA's Small Business Innovation Research Program, grants awarded under this topic area develop new technologies or apply existing technologies to address concerns facing rural America. Efforts focus on environmental enhancement, disaster resilience, service delivery, and entrepreneurial and workforce skills. Topics may include technologies and services that promote rural tourism, protect the ecosystem, conserve energy, and develop alternative energy sources such as wind and solar energy (excluding biofuels).	The Small Business Innovation Research program has been reauthorized through September30, 2017.
Small Business Innovation Research Program: Small and Mid-Size Farms Topic Area	As part of USDA's Small Business Innovation Research Program, grants awarded under this topic area develop new technologies and information that will help improve the viability and profitability of small and mid-sized farms and ranches.	The Small Business Innovation Research program has been reauthorized through September 30, 2017.

Table 5. (Continued)

Initiative name and implementing subagency	Description	Expiration information
	Efforts include developing enterprises focused on specialty farm products, enhancing the efficiency and profitability of small farms, developing farming methods for small farms that use natural resources more efficiently, and developing new educational tools to ensure that small farmers have the information they need to operate their farms on a sustainable and profitable basis. Particular emphasis is placed on research to utilize renewable energy sources, such as wind, solar, and geothermal energy, and to promote improved energy efficiency and conservation in farming operations.	
Natural Resources Conservation Service		
Conservation Innovation Grant Program	This program, authorized under the Food, Conservation and Energy Act of 2008, provides grants to tribes, state or local governments, nongovernmental organizations, or individuals to develop, evaluate, implement, and monitor conservation approaches, incentive systems, or technologies. The purpose of these grants is to stimulate the development and adoption of innovative conservation approaches and technologies. The Natural Resources Conservation Service expects these projects to lead to the transfer of conservation technologies, systems, and approaches into agency policy, technical manuals, and guidance, or to the private sector.	Authorization for this program expires with the end of the Food, Conservation, and Energy Act of 2008. However, the Department ofAgriculture reported that, should funding remain available after the expiration of the act, it has the authority to continue program operations based upon existing regulations.

Initiative name and implementing subagency	Description	Expiration information
	Among other areas of focus, the Natural Resources Conservation Service awards grants for projects that focus on innovative tools to estimate the energy and fossil fuel implications of cropland and conservation practices; innovative on-farm applications of renewable energy production technologies; and sustainable biomass production, harvesting, and handling technologies.	
Conservation Security Program	This program was not reauthorized in the 2008 Farm Bill and is no longer available; however, contracts written while the program was still authorized can span up to 10 years and some contracts were still being implemented in fiscal year 2011. This provided financial and technical assistance to promote the conservation and improvement of soil, water, air, energy, plant, and animal life, and other conservation purposes on tribal and private lands. The Natural Resources Conservation Service established contracts for two types of renewable energy-related efforts: (1) payments for the bio-based portion of eligible blended fuel; and (2) payments for electricity generated from renewable sources, including wind energy.	This program no longer exists.
Conservation Stewardship Program	This program, authorized under the Food, Conservation and Energy Act of 2008, provides financial and technical assistance to help agricultural and forestry producers conserve and enhance soil, water, air, and related natural resources on their land by (1) undertaking additional conservation activities and	None

Table 5. (Continued)

Initiative name and implementing subagency	Description	Expiration information
	(2) improving and maintaining existing conservation systems. One such conservation activity that is eligible for assistance under the program is for pumping plants powered by renewable energy—wind or solar—to supply water for irrigation, drainage, livestock watering, or wildlife.	
Environmental Quality Incentives Program	This program, authorized under the Food, Conservation and Energy Act of 2008, provides financial and technical assistance through contracts with agricultural producers to implement conservation practices to address environmental natural resource problems. The Natural Resources Conservation Service reported that, while the program does not fund projects for the sole purpose of energy production, in cases of remote livestock where no source of electricity is available, eligible producers may pursue wind or solar power to support the pumping of water for livestock in remote regions of rangeland and where multiple resource concerns are addressed.	Authorization for this program expires with the end of the Food, Conservation, and Energy Act of 2008. However, the Department ofAgriculture reported that, should funding remain available after the expiration of the act, it has the authority to continue program operations based upon existing regulations.
Office of the Chief Economist		
Energy and Bioenergy Research	The Office of the Chief Economist advises the Secretary of Agriculture on economic issues related to agriculture, such as commodity trends and issues, as well as specific topics like renewable energy. The Office of the Chief Economist's Office of Energy Policy and New Uses assists the Secretary of Agriculture in developing and coordinating USDA energy policy, programs, and strategies.	None

Initiative name and implementing subagency	Description	Expiration information
	The mission of this research program is to conduct economic analyses and evaluate policies and strategies concerning biomass feedstocks for renewable energy production. Efforts under this program have largely focused on liquid transportation fuels and energy sources used to produce heat and power, including wind.	
Office of the Secretary		
USDA / Navy Memorandum of Understanding Project	Through this effort, USDA is working with the Navy to reduce energy consumption derived from fossil fuels, increase clean energy production from renewable energy sources to meet transportation needs, help meet the Secretary of the Navy's goal of sailing the "Great Green Fleet," and to support the President's "Blueprint for a Secure Energy Future." To accomplish this, USDA is engaging other departmental programs. USDA has established a memorandum of understanding with the Navy in support of this effort under which the departments agreed to share technical, program management, and financial expertise, and to cooperate in developing strategies and plans to implement renewable energy projects and initiatives. USDA and the Navy also agreed to collaborate on the funding of projects, and to support development of advanced biofuels and renewable energy processes that are sustainable from an economic, social, and environmental perspective. Wind energy would have been eligible for support through this initiative in fiscal year 2011, in that the initiative complements other USDA efforts that support wind energy, such as the Rural	This initiative is implemented through other programs that wereset to expire at the end of the Food, Conservation, and Energy Act of 2008. Many programs under that act were extended by the American Taxpayer Relief Actof 2012, including biofuel and other renewable energy provisions.

Table 5. (Continued)

Initiative name and implementing subagency	Description	Expiration information
	Energy for America Program, as well as the Navy and Marine Corps' existing renewable energy programs. However, the initiative did not fund any wind projects in 2011, and no future funding is planned for wind projects.	
Rural Business-Cooperative Service		
Business and Industry Guaranteed Loan Program	The purpose of this program is to improve, develop, or finance business, industry, and employment, and the economic and environmental climate in rural communities by providing guaranteed loans to borrowers in rural areas. Eligible project proposals include those that will provide employment; improve the economic or environmental climate; promote the conservation, development, and use of water for aquaculture; or reduce reliance on nonrenewable energy resources by encouraging the development and construction of renewable energy systems, including wind energy systems. Loan amounts are generally limited to $10 million per borrower except under certain circumstances, and the percentage amount of the loan guarantee varies depending upon the value of the loan.	None
Rural Energy for America Program (formerly the Renewable Energy Systems and Energy Efficiency Improvements Program)	This program, authorized under the Food, Conservation and Energy Act of 2008, provides funding for guaranteed loans and grants to farmers, ranchers, and small businesses in rural areas to assist with purchasing and installing renewable energy systems, such as wind energy projects, and energy efficiency improvements.	Authorization for this program wasset to expire with the end of the Food, Conservation, and Energy Act of 2008. On January 2, 2013, the American Taxpayer Relief Act of 2012 authorized

Initiative name and implementing subagency	Description	Expiration information
	Guaranteed loans under the program encourage the commercial financing of renewable energy and energy efficiency projects by guaranteeing between 60 and 85 percent of the loan (depending upon the amount of the loan). Loans may be guaranteed up to 75 percent of a project's cost or $25 million. Grants under the program are awarded on a competitive basis and can be up to 25 percent of total eligible project costs or $500,000 for renewable energy systems and $50,000 for renewable energy feasibility studies. At least 20 percent of the renewable energy system grant funds awarded must be for grants of $20,000 or less. In addition, the program provides grants for energy audits.	discretionary funding for the program for fiscal year 2013. The Department of Agriculture reported that, should funding remain available after the expiration of the act, it has the authority to continue program operations based upon existing regulations.
Rural Utilities Service		
Direct and Guaranteed Electric Loan Program	This program, authorized under the Rural Electrification Act of 1936, provides loans and loan guarantees to establish and improve electric service in rural areas and to assist electric borrowers to implement demand side management, energy efficiency and conservation programs, and on-grid and off-grid renewable energy systems including wind energy systems. These loans and loan guarantees provide financing to eligible nonprofit utility organizations, such as electric co-ops and public utility districts, as well as for-profit entities. The Rural Utilities Service reported that providing loans for renewable energy is one of the primary purposes of the Rural Electrification Act and, to advance this purpose, the agency gives priority to completed renewable energy project	None

Table 5. (Continued)

Initiative name and implementing subagency	Description	Expiration information
	loan applications in its annual processing queue. In addition, Rural Utilities Service reported that it is reviewing its policies and regulations to initiate, as appropriate, additional accommodations to support renewable energy project lending to a wider range of applicants consistent with the underlying statutory requirements for prudent lending and adequate loan security.	
High Energy Cost Grant Program	This program, authorized under the Rural Electrification Act of 1936, provides grants for energy generation, transmission, and distribution facilities serving rural communities with annual average home energy costs that exceed 275 percent of the national average. Applicants may receive grants for on-grid and off-grid renewable energy systems, such as wind energy systems, as well as energy conservation and efficiency projects. In addition, the Rural Utilities Service reported that the Rural Electrification Act also includes loan authority, but that no funds have been appropriated for loans under this program.	None

Source: GAO analysis of agency-provided data.

Table 6. Department of Commerce Wind-Related Initiatives

Initiative name and implementing subagency	Description	Expiration information
Economic Development Administration		
Environmentally-Sustainable Development Investment Priority	Economic Development Administration awards grants that are intended to advance the administration's key investment	None

Initiative name and implementing subagency	Description	Expiration information
	priorities, one of which is environmentally sustainable development. This priority focuses the Economic Development Administration's existing place-based grant programs (planning, technical assistance, and infrastructure construction) in support of promoting economic development projects that, among other things, enhance environmental quality. Eligible projects include wind and other renewable energy economic development projects.	
Global Climate Change Mitigation Incentive Fund	The Global Climate Change Mitigation Incentive Fund was established to strengthen the linkages between economic development and environmental quality. It finances economic development projects that create jobs and increase private capital investment in efforts to limit the nation's dependence on fossil fuels, enhance energy efficiency, curb greenhouse gas emissions, and protect natural systems. Eligible projects include economic development projects that support wind and other sources of renewable energy.	The Global Climate Change Mitigation Incentive Fund is not astanding program, but rather a congressionally directed funding stream, which requires annual submission of a report/spending plan to Congress summarizing how the funds will be used.
International Trade Association		
International Buyer Program	The International Buyer Program is an initiative that supports U.S. companies at U.S. trade shows by recruiting and escorting foreign buyer delegations to the shows to meet with U.S. companies to participate in business-to-business matchmaking activities. These shows often include one-on-one meetings between U.S. companies and the international buyers.	None

Table 6. (Continued)

Initiative name and implementing subagency	Description	Expiration information
	The initiative supported several major shows focused on renewable energy, including the WINDPOWER 2012 Conference & Exhibition featuring 51 foreign buyer delegates.	
Market Development Cooperator Program	Under this initiative, industry groups—such as trade associations and chambers of commerce—compete for International Trade Association-awarded grants that fund projects to enhance their industry's competitive position in innovative ways. The selected groups establish partnerships with the International Trade Association, which provides financial and technical assistance to enhance global competitiveness of U.S. industries. Partnerships have included several related to renewable energy, such as the Green Export Enabler program and the Colorado Export of Innovative and Sustainable Technologies program.	None
Renewable Energy and Energy Efficiency Export Initiative	This initiative represents the first-ever national effort to coordinate export promotion of renewable energy and energy efficiency technologies, including wind energy technologies. It is led by the Department of Commerce, and includes commitments from eight U.S. government agencies—including the Departments of Agriculture, Energy, and State—to help tailor financing products to the needs of U.S. renewable energy and energy efficiency companies, enhance market access and trade promotion for their products, and	None

Initiative name and implementing subagency	Description	Expiration information
	improve the delivery of U.S. government services to these companies. An example of projects supported by the initiative includes the commitment to lead several trade policy missions to key potential markets for U.S. companies.	
National Institute of Standards and Technology		
National Institute of Standards and Technology Smart Grid Program	The National Institute of Standards and Technology's efforts in this area involve coordinating the development of interoperability standards needed to accelerate implementation of a smart grid system. The smart grid is a planned nationwide electric transmission and distribution network that uses information technology to deliver electricity efficiently, reliably, and securely. Conceptually, a smart grid system will enable integration of wind and other renewable and alternate energy sources into the electric grid.	None
National Oceanic and Atmospheric Administration		
Joint Wind Energy Program: Atmospheric Velocity Gradients	Through a cooperative agreement, the National Oceanic and Atmospheric Administration's Air Research Laboratory is working with the private sector to conduct wind energy research to demonstrate the potential for significantly improving the prediction of winds at various heights. These efforts will help improve short-term predictions of wind farm energy production.	None
Program Development	This initiative explores ways to leverage the National Oceanic and Atmospheric Administration Earth Systems Research Laboratory's ongoing research for renewable energy purposes.	None

Table 6. (Continued)

Initiative name and implementing subagency	Description	Expiration information
	For example, while another National Oceanic and Atmospheric Administration program collects wind data for atmospheric research, this initiative explores how that same data could be used for renewable energy purposes.	
Renewable Energy Research	The purpose of this initiative is to conduct research in four areas: (1) wind power observations and forecasting, (2) solar power observations and forecasting, (3) climate analyses to assess wind and solar resources, and (4) climate change impacts on wind and solar resources. For instance, a study has been under way to determine wind and solar resource availability over the continental United States, and a comparison of the estimated wind and solar power that could be produced against energy load during the years 2006 through 2008. Researchers will present their results to academia, industry, and others.	None
MarineCadastre.gov	The purpose of this program is to support wind energy development and other uses of the outer continental shelf by providing mapping information on locations for project planning and siting. This information is intended to help renewable energy developers in initial planning stages to identify multiple uses and avoid conflicts before producing plans for development. This initiative is managed jointly by the National Oceanic and Atmospheric Administration and the Department of the Interior's Bureau of Ocean	None

Initiative name and implementing subagency	Description	Expiration information
	Energy Management, Regulation and Enforcement.	
U.S. Patent and Trademark Organization		
Green Technology Pilot Program	This initiative supports investment in renewable energy, including wind energy, by spurring inventive activity with the prospect of early intellectual property protection. Such protection is often critical in the ability to raise the venture capital for development and commercialization of the technology. The U.S. Patent and Trademark Organization's Patent Office reviews applications pertaining to green technologies—such as development of renewable energy resources—on an expedited basis, which allows for earlier acquisition of patent protection.	The Green Technology Pilot Program expired in February2012.

Source: GAO analysis of agency-provided data.

Table 7. Department of Energy (DOE) Wind-Related Initiatives

Initiative name and implementing subagency	Description	Expiration information
Advanced Research Projects Agency – Energy		
Agile Delivery of Electrical Power Technology	This initiative supports Advanced Research Projects Agency-Energy's efforts to invest in materials—such those used in power converters—that facilitate key advances in electrical power technology, including those that enable solar photovoltaics and wind turbine generators.	The Advanced Research Projects Agency-Energy's authorizing legislation, the America Creating Opportunities to Meaningfully Promote Excellence in Technology, Education, and Science Act of 2010, was recently reauthorized until 2013.

Table 7. (Continued)

Initiative name and implementing subagency	Description	Expiration information
Advanced Research Projects Agency-Energy Funding Opportunity Announcement 1	Under this initiative, Advanced Research Projects Agency-Energy provides funding to support research on a variety of energy ideas and technologies. This research funding is focused on applicants with well-formed research and development plans for potentially high-impact concepts or new technologies, including wind and other renewable technologies.	The Advanced Research Projects Agency-Energy's authorizing legislation, the America Creating Opportunities to Meaningfully Promote Excellence in Technology, Education, and Science Act of 2010, was recently reauthorized until 2013.
Grid-Scale Rampable Intermittent Dispatchable Storage	This initiative supports Advanced Research Projects Agency-Energy's efforts to develop new technologies that enable widespread use of cost-effective grid-scale energy storage. Specifically, it focuses on developing technologies that can be widely implemented across the power grid to mitigate variability in energy generated from renewable sources such as wind.	The Advanced Research Projects Agency-Energy's authorizing legislation, the America Creating Opportunities to Meaningfully Promote Excellence in Technology, Education, and Science Act of 2010, was recently reauthorized until 2013.
Office of Energy Efficiency and Renewable Energy (EERE)		
Energy Efficiency and Conservation Block Grants	The Energy Efficiency and Conservation Block Grant Program provides assistance to cities, counties, states, territories, and tribes to develop and implement projects and programs that reduce fossil fuel emissions; reduce the total energy use of the eligible entities;	This program was funded through American Recovery and Reinvestment Act of 2009 funding,and no additional appropriations were granted. The Recovery Act funds were required to be obligated by September 30, 2010,

Initiative name and implementing subagency	Description	Expiration information
	improve energy efficiency in various economic sectors; and create and retain jobs. Most of the funding provided under this program supports formula grants, although some funding supports competitively awarded grants or is used to develop technical assistance tools. Program funds may be used for a variety of energy efficiency and conservation programs and projects, as well as for renewable energy projects including wind projects. Renewable energy projects that generate electricity and are installed on or in a government building are expressly eligible under the statute. Other renewable energy activities may be eligible as part of a comprehensive energy efficiency program. DOE reported that a significant number of grantees have used program funding to implement projects that include renewable energy elements.	and expended by September 30, 2013.
Hydrogen and Fuel Cell Technologies Crosscutting Activities	This initiative focuses on crosscutting functions for hydrogen and fuel cell technologies such as manufacturing, technology validation, safety codes and standards, and education. It also analyzes technologies to identify gaps and help direct future research and development.	The Energy Policy Act of 2005 and the Energy Independence andSecurity Act of 2007 provide authorization for these activities through fiscal year 2015 or 2020, depending upon the activity.

Table 7. (Continued)

Initiative name and implementing subagency	Description	Expiration information
	Hydrogen and fuel cell technologies include those produced from renewable resources such as biomass, solar, wind, etc.	
Hydrogen Fuel R&D	This initiative focuses on research and development to develop hydrogen production, delivery, and storage technologies, including those from hydrogen produced using renewable energy sources such as biomass, solar, wind, etc.	The Energy Policy Act of 2005 and the Energy Independence andSecurity Act of 2007 provide authorization for these activities through fiscal year 2015 or 2020, depending upon the activity.
State Energy Program	The State Energy Program provides financial and technical assistance to grantees through formula and competitive grants for the deployment of energy efficiency and renewable energy technologies, including wind energy technologies. States use their formula grants, plus a 20 percent match that they provide, to develop state strategies and goals to address their energy priorities. In designing their programs, grantees choose a list of eligible activities to fund with their grant, including a variety of renewable energy-related activities. DOE reported that a significant number of grantees have incorporated renewable energy activities into their state energy	None

Initiative name and implementing subagency	Description	Expiration information
	programs. Competitive grants awarded under the State Energy Program assist grantees with the adoption of energy efficiency/renewable energy products and technologies.	
Tribal Energy Program	This program promotes energy self-sufficiency, economic development, and employment on Tribal lands through the use of renewable energy and energy efficiency technologies, including wind energy technologies. Under this program, DOE provides competitively-awarded funding and technical assistance to Tribes to evaluate and develop their renewable energy resources and reduce their energy consumption through efficiency and weatherization. In addition, DOE provides training and outreach to Tribal governments, including student internships; regional workshops and Webinars to learn about renewable energy and energy efficiency and how to develop those resources; and renewable energy short courses covering renewable energy and efficiency options, business development, and project financing.	None

Table 7. (Continued)

Initiative name and implementing subagency	Description	Expiration information
Wind Energy - Offshore Wind	Through this initiative, EERE seeks to promote and accelerate the deployment of offshore wind power technologies in U.S. waters. Under this initiative, DOE plans to develop innovative technologies to reduce the cost of offshore wind energy. DOE will work with Interior and other agencies to shorten the timeline for deploying these technologies, and will partner with industry to demonstrate commercially and technically viable offshore wind technology. Efforts include research, development, demonstration, education, outreach, and interagency coordination.	None
Wind Energy - Technology Application	Under this initiative, EERE helps prepare and accelerate the market adoption of wind technologies. It includes two components: (1) systems integration and (2) technology acceptance. The systems integration component focuses on planning and operating wind energy projects, and anticipating and overcoming technical issues associated with interconnecting large amounts of wind and other renewable energy to the electricity grid.	None

Initiative name and implementing subagency	Description	Expiration information
	The technology acceptance component addresses opportunities for and barriers to the use of wind energy systems such as mitigating wind-radar interference; providing state and regional energy sector education and outreach; and investigating and mitigating social, environmental, and wildlife issues associated with wind energy development.	
Wind Energy - Technology Viability	This initiative's activities focus on research, development, and testing to improve the performance, cost-effectiveness, manufacturing, and reliability of large and distributed wind energy systems. It has three components: (1) low wind speed technology, (2) distributed wind technology, and (3) supporting research and testing. Distributed wind technology, which leverages transmission availability off the main grid, enables community ownership and builds support for wind in general.	None
Loan Programs Office		
Title XVII Section 1703 Loan Guarantee Program	Under Section 1703 of Title XVII of the Energy Policy Act of 2005, DOE provides loans to support innovative energy technologies that are typically unable to obtain conventional private financing due to high technology risks.	None

Table 7. (Continued)

Initiative name and implementing subagency	Description	Expiration information
	In addition, the technologies must avoid, reduce, or sequester air pollutants or emissions of greenhouse gases. Eligible technologies under this initiative included biomass, solar, wind, hydropower, and alternative fuel vehicles, among others. Under Section 1703, borrowers were to pay DOE for the credit subsidy costs. However, DOE noted that once the Section 1705 Loan Guarantee Program (under which the credit subsidy cost is paid with appropriated funds) was established under the Recovery Act specifically to support renewable energy and certain other projects, renewable energy projects were supported under the Section 1705 Loan Guarantee Program rather than the Section 1703 Loan Guarantee Program. DOE now has $170 million in appropriated funds to pay the credit subsidy cost for renewable energy and energy efficiency projects under Section 1703.	
Title XVII Section 1705 Loan Guarantee Program	The Recovery Act added Section 1705 to the Energy Policy Act of 2005. Section 1705 authorized a temporary program to provide loan guarantees for certain renewable energy systems,	The American Recovery and Reinvestment Act of 2009 added Section 1705 to the Energy PolicyAct of 2005.

Initiative name and implementing subagency	Description	Expiration information
	electric power transmission systems and innovative biofuel projects that began construction no later than September 30, 2011. To help implement this authority, DOE established the Financial Institution Partnership Program, a risk-sharing partnership between DOE and qualified finance organizations. Through this program, DOE pays the credit subsidy costs of loan guarantees using funds appropriated for this purpose, and guarantees up to 80 percent of a loan provided for a renewable energy generation project by qualified financial institutions. In addition, under the 1705 authority, DOE funded eligible innovative wind and other renewable energy projects through the Federal Financing Bank.	The initiative expired on September 30, 2011.
Multiple		
Small Business Innovation Research/Small Business Technology Transfer - Wind Energy Technology Development Topic Area	Under this initiative, DOE provides grant funding to stimulate technological innovation in small businesses to meet federal agency research and development needs. The solicitation contains technical topics in a number of research areas of interest to DOE. Grants awarded under this topic area promote the development of wind energy technologies.	Continuation of the Small Business Innovation Research/Small Business Technology Transfer programs is contingent upon their reauthorization.

Table 7. (Continued)

Initiative name and implementing subagency	Description	Expiration information
	Efforts focus on manufacturing and assembly, component reliability, and monitoring.	
Office of Electricity Delivery & Energy Reliability		
Clean Energy Transmission and Reliability	This initiative focuses on developing tools and techniques to address system reliability issues, including challenges that renewable energy sources such as wind and solar power pose to electricity grid operators, who need to incorporate these variable generation sources without compromising reliability. The Office of Electricity Delivery & Energy Reliability's efforts under this initiative also include communications, education, and outreach.	None
Funding of five interconnection-wide transmission planning & associated projects	Through this initiative, the Office of Electricity Delivery & Energy Reliability aims to facilitate the development or strengthening of capabilities in each of the three interconnections (Western, Eastern, Texas) serving the lower 48 states of the United States by preparing analyses of transmission requirements under a broad range of alternative futures and developing long-term interconnection-wide transmission expansion plans.	None

Initiative name and implementing subagency	Description	Expiration information
	This includes efforts to identify renewable resources such as wind that are suitable for near-term development to supply electricity to meet state renewable portfolio standards.	
Office of Electricity Delivery & Energy Reliability Energy Storage	The goal of this initiative is to develop advanced energy storage technologies and systems that will increase the reliability, performance, and competitiveness of electricity generation and transmission in the electric grid and in stand-alone systems. These efforts also facilitate integration of renewable energy sources such as wind into the electricity grid. Through the initiative, DOE works to build a community around this goal in collaboration with industry, academia, and government institutions to ensure that technology development meets all stakeholder needs and interests. The initiative also provides grants to private companies to encourage commercialization and deployment.	None
Office of Electricity Delivery & Energy Reliability Permitting, Siting, and Analysis – various projects	This initiative provides technical and financial assistance to states and regional entities to help develop infrastructure required to utilize renewable energy, among other resources.	None

Table 7. (Continued)

Initiative name and implementing subagency	Description	Expiration information
	For example, the initiative assists state public utility commissions, state legislatures, regional state associations, and governors' offices on renewable energy policies and portfolio standards, among other topics. Among other things, DOE issued its second electric transmission congestion study, which includes consideration of remote renewable resources including wind to comply with provisions of the Recovery Act.	
Office of Indian Energy Policy and Programs		
Strategic Technical Assistance Response Team Program	The Strategic Technical Assistance Response Team initiative is a project of the DOE Office of Indian Energy Policy and Programs aimed at advancing next-generation energy development in Indian Country. Through the Strategic Technical Assistance Response Team Program, early-stage project development technical assistance is provided to 10 selected projects—5 in Alaska and 5 in the 48 contiguous states.	None
Power Marketing Administrations		
Bonneville Power Administration Wind Integration	The Bonneville Power Administration is a federal entity that is also one of the largest transmission providers in the West.	None

Initiative name and implementing subagency	Description	Expiration information
	The Bonneville Power Administration has played an active role in facilitating the development of wind energy in the Pacific Northwest, and its area of operation has become a test bed for management of large and growing amounts of intermittent wind energy to peak electricity load. The Bonneville Power Administration is piloting and implementing operational practices to manage the intermittent wind resource in a variety of key areas. The Bonneville Power Administration works to help the region meet its energy and environmental goals while simultaneously providing reliable load service and satisfying multiple nonpower constraints, such as protection of endangered fish.	
Western Area Power Administration—Operations	Western Area Power is participating in the Western Interconnection Joint Initiatives projects intended to facilitate the integration of variable energy resources, including wind energy. These initiatives include tools or processes such as Intra-hour Scheduling, Dynamic Scheduling System, and Intra-hour Transaction Accelerator Platform.	None

Table 7. (Continued)

Initiative name and implementing subagency	Description	Expiration information
Western Area Power Administration—Transmission Services	The Western Area Power Administration began its Transmission Infrastructure Program in 2009 under section 402 of the Recovery Act, for the purpose of constructing, financing, facilitating, planning, operating, maintaining, or studying construction of new or upgraded electric power transmission lines and related facilities with at least one terminus within the area served by Western, and for delivering or facilitating the delivery of power generated by renewable energy resources constructed or reasonably expected to be constructed after the date section 402 was enacted. The program uses authority granted under this section to borrow up to $3.25 billion from the U.S. Department of the Treasury to accomplish these purposes. Pursuant to the Recovery Act, projects under the program must (1) be in the public interest; (2) not adversely impact system reliability or operations, or other statutory obligations; (3) offer reasonable expectation that proceeds will be adequate to meet Western's repayment obligations; (4) use a public process to set transmission rates;	None

Initiative name and implementing subagency	Description	Expiration information
	(5) have the necessary capability to obtain and deliver generation-related ancillary services; and (6) provide proceeds to repay principal and interest of the loan from the Treasury.	

Source: GAO analysis of agency-provided data.

Table 8. Department of the Interior Wind-Related Initiatives

Initiative name and implementing subagency	Description	Expiration information
Bureau of Land Management		
Recovery Act Renewable Energy Efforts	These efforts are intended to support the Bureau of Land Management's portfolio of renewable energy work, through studies related to wind and solar energy right-of-way authorizations on the bureau's lands. Projects funded under the Recovery Act included resource assessments and studies on environmental impacts of wind and solar development.	These are one-time activities authorized under the American Recovery and Reinvestment Actof 2009.
Renewable Energy Coordination Offices Implementation	To facilitate the development of wind and solar energy projects on Bureau of Land Management lands, the bureau has increased dedicated teams of staff to streamline the review and approval of applications and conduct analysis required by the National Environmental Policy Act. In particular, this initiative facilitated further development of the Renewable Energy Coordination Offices through staffing of the Renewable Energy Support personnel in Colorado, Idaho, Montana, New Mexico, Oregon/Washington, and Utah.	None

Table 8. (Continued)

Initiative name and implementing subagency	Description	Expiration information
	Support personnel provide additional capacity and resources in those states with lower volume workload but significant renewable energy workloads.	
Wind Energy Authorizations and Operations on Bureau of Land Management Public Lands	The purpose of this initiative is to regulate wind energy development on Bureau of Land Management lands. The bureau processes and approves developers' applications, and collects land rents and other fees as projects are built and operated.	None
Bureau of Ocean Energy Management, Regulation and Enforcement (BOEMRE)a		
Environmental Studies Program	Through this program, BOEMRE provides information needed for prediction, assessment, and management of impacts on the human, marine, and coastal environments of the outer continental shelf and nearshore areas that may be affected by oil and gas, wind and other renewable energy, or other development. Specific research activities focus on establishing baseline information for assessing environmental impacts, constructing models to predict impacts, and environmental monitoring to provide trend data on significant changes in the environment.	None
Renewable Energy Program Development and Implementation	Through these efforts, BOEMRE implements and manages its program to authorize orderly, safe, and environmentally responsible renewable energy development on the outer continental shelf, as established in 2009 by the department's renewable energy regulatory framework document.	None

Initiative name and implementing subagency	Description	Expiration information
	The bureau also coordinates these efforts with ongoing state and local renewable energy efforts, as required by the Energy Policy Act of 2005. Specific efforts include renewable energy resource assessment and transmission studies; environmental studies; public stakeholder meetings; and support for intergovernmental and interagency meetings and information exchange. These efforts support the bureau's efforts to grant leases, easements, and rights-of-way for wind and other renewable energy development activities on the outer continental shelf.	
Renewable Energy Program: Ensure Fair Return for Renewable Energy Resources	By statute, renewable energy leases must be issued competitively unless the Secretary determines, after public notice of a proposal, that there is no competitive interest. BOEMRE is responsible for ensuring the public will receive a fair return for the competitive and noncompetitive award of renewable energy leases. The regulations authorize a range of auction formats that have not been used previously by BOEMRE. These formats may be implemented under appropriate circumstances to encourage competitive auctions. BOEMRE's ability to set reasonable commercial lease terms and auction parameters necessary to earn a fair return is directly related to its ability to conduct accurate wind resource assessments and economic modeling of projects within a given wind energy area. The initiative supports BOEMRE on these issues.	None

Table 8. (Continued)

Initiative name and implementing subagency	Description	Expiration information
Renewable Energy Program: Environmental Compliance	Through this program, BOEMRE works to ensure that the bureau's renewable energy activities comply with all applicable laws and regulations, including the National Environmental Policy Act. To facilitate siting, leasing, and construction of new projects, the bureau prepares environmental assessments and environmental impact statements to inform decisions on lease and grant issuance, and plan approval. The bureau has focused these efforts on supporting the Secretary of the Interior's "Smart form the Start" initiative to facilitate the development of wind energy on the Atlantic Outer Continental Shelf.	None
Renewable Energy Program: Multipurpose Marine Cadastre	This program supports wind energy development, and coastal and marine spatial planning on the outer continental shelf in general, by providing mapping information on locations for project planning and siting. The bureau provides data, data services, and a geospatial viewer with information on state and federal areas of jurisdiction over marine and coastal property, as well as other nonjurisdictional data sets. This information is intended to help renewable energy developers in initial planning stages to identify multiple uses and avoid conflicts before producing plans for development. The National Oceanic and Atmospheric Administration provides in-kind support as the technical lead agency for this program.	Authority for this initiative will expire at the end of fiscal year2014.

Initiative name and implementing subagency	Description	Expiration information
Renewable Energy Program: Safety Program and Inspections	The current focus of this program is to identify the critical components of an offshore wind turbine, how it might fail, and what inspection techniques should be followed. The bureau has conducted studies to identify this information and used it to develop detailed guidance for agency inspectors for the inspection of offshore wind energy installations. The office has also commissioned a study by the National Academy of Science to determine best practices for the design, fabrication, and installation of offshore wind turbines, and is collaborating with the American Wind Energy Association to develop a consensus industry guideline on appropriate standards for these turbines.	None
Resource Evaluation Program: Economic Analysis	BOEMRE has a responsibility to conduct economic analyses for the evaluation of resources on tracts offered for development, to ensure the public receives a fair return. BOEMRE conducts financial analysis of the fees to be included in commercial leases, provides semiannual renewable energy revenue and activity projections to the Office of Management and Budget, and conducts benefit-cost analysis of proposed legislative and regulatory policies, which will also address regional economic assessments as the program evolves.	None
Resource Evaluation Program: Resource Evaluation	Resource evaluation is necessary to ensure a fair return and to assess the offshore renewable energy resource potential in areas that are being considered for leasing.	None

Table 8. (Continued)

Initiative name and implementing subagency	Description	Expiration information
	Presently, this initiative is primarily focused on the collection and quantification of wind resource data developed by the DOE, as well as the development of in-house capabilities with respect to the collection and quantification of wind, current, and wave resources. Other objectives of this initiative include use of the information by BOEMRE to determine minimum bid amounts for competitive lease sales, providing review of potentially dangerous conditions related to the placement of facilities on the seabed, and generating other information necessary to support lease sales.	
Technology Assessment and Research Program	Under this program, the bureau seeks to ensure that industry operations on the outer continental shelf incorporate the use of the best available and safest technologies, as required by federal laws including the Outer Continental Shelf Lands Act Amendments of 1978 and the Energy Policy Act of 2005. There are two components to the Technology Assessment and Research Studies Program: operational safety and engineering research, and renewable energy research. Within the renewable energy research component, the office has funded industry researchers to study offshore wind technologies.	None

Initiative name and implementing subagency	Description	Expiration information
Bureau of Reclamation		
Desalination and Water Purification Research and Development Program	This program is intended to address the needs for water desalination and purification in locations that do not have adequate traditional sources of water by lowering the costs of desalination and purification technologies and understanding and minimizing associated environmental impacts. Through the program, the Bureau of Reclamation provides financial assistance to water utilities, state and local governments, industry, and other parties, and emphasizes the application of renewable energy technologies such as wind, particularly for minimizing environmental impacts of desalination and purification.	This initiative ended at the end offiscal year 2011.
Science and Technology Program	This program provides funding on a competitive basis to Bureau of Reclamation researchers to improve the operations, efficiency, and maintenance of the bureau's hydropower resources. The bureau also provides funding to support research on other renewable energy generation systems, such as wind systems, to understand how they can improve the bureau's ability to carry out its mission. This office also funds an interagency agreement with the National Renewable Energy Laboratory to assist Reclamation in developing both facility- and utility-scale renewable installations.	None
Bureau of Indian Affairs		
Minerals & Mining Program: Renewable Energy Projects	This program funds a broad range of renewable energy projects on tribal lands, with the goal of enabling tribes to take over the	None

Table 8. (Continued)

Initiative name and implementing subagency	Description	Expiration information
	federal government's role of oversight and development of resources on tribal lands. Indian Affairs funds projects through an annual competitive grant award process, and projects have included economic analyses, feasibility studies, resource assessments, marketing of resources, and construction of facilities. These projects have involved a wide range of renewable resources, including woody biomass, waste conversion, geothermal, hydroelectric, solar, and wind.	
Office of Insular Affairs		
Insular Plan for Alternative and Renewable Energy	As part of a broader strategic planning effort for the islands required by the Assistant Secretary of Insular Affairs, the Governors of the Pacific territories of American Samoa, Guam, and the Commonwealth of the Northern Mariana Islands each established energy steering committees to guide the development of energy on their islands. In addition, the Office of Insular Affairs funds DOE's National Renewable Energy Laboratory to conduct renewable energy assessments and develop draft strategic energy plans for each territory. Among other things, these plans have identified opportunities for the territories to develop wind, solar, geothermal, biomass, power system upgrades, improvements to existing diesel generators, and the implementation of island-wide energy efficiency and conservation programs.	None

Initiative name and implementing subagency	Description	Expiration information
	The Office of Insular Affairs recently expanded the laboratory's scope of work to include addressing the needs of the U.S. freely associated states, which include the Republic of the Marshall Islands, the Republic of Palau, and the Federated States of Micronesia. Working with the island leadership, local utilities, and other stake holders, the laboratory is examining realistic, cost-effective ways of employing renewable energy to solve the biggest economic challenge facing the islands, i.e., their 100 percent dependence on imported fossil fuels.	

Source: GAO analysis of agency-provided data.

[a] On October 1, 2011, BOEMRE reorganized into two independent entities: the Bureau of Ocean Energy Management and the Bureau of Safety and Environmental Enforcement. The Bureau of Ocean Energy Management is responsible for managing development of the nation's offshore resources in an environmentally and economically responsible way, and its activities include oversight of leasing, environmental studies, and economic analysis. The Bureau of Safety and Environmental Enforcement is responsible for enforcing safety and environmental regulations. However, given the time frame of our work, we refer to BOEMRE in this report.

Table 9. Department of the Treasury Wind-Related Initiatives

Initiative name and implementing subagency	Description	Expiration information
Internal Revenue Service (IRS)		
Accelerated Depreciation Recovery Periods for Specific Energy Property	Under the federal Modified Accelerated Cost-Recovery System, businesses may recover the costs of investments in certain property through depreciation deductions from a taxpayer's taxable income.	Bonus depreciation was extended by the American Taxpayer Relief Act of 2012 through 2013 for qualifying property placed in service before January 1, 2014

Table 9. (Continued)

Initiative name and implementing subagency	Description	Expiration information
	The system sets the period of time over which various types of property may be depreciated from 3 to 50 years, and classifies a number of renewable energy technologies as 5-year property, including wind. The Joint Committee on Taxation generally classifies as tax expenditures cost recovery allowances that are more favorable than those provided under the alternative depreciation system (Internal Revenue Code Section 168(g)), which provides for straight-line recovery over tax lives that are longer than those permitted under the accelerated system. However, the Economic Stimulus Act of 2008 included a 50 percent first-year bonus depreciation provision for a wide range of eligible properties including renewable energy systems. This provision was extended by the American Recovery and Reinvestment Act of 2009, and by the Creating Small Business Jobs Act of 2010. Bonus depreciation was further extended through 2012 by the Tax Relief, Unemployment Insurance Reauthorization, and Job Creation Act of 2010, with a 100 percent deduction allowed for property acquired after September 8, 2010, and before January 1, 2012.	(or January 1, 2015 for certain other assets).

Initiative name and implementing subagency	Description	Expiration information
	The American Taxpayer Relief Act of 2012 extended 50 percent expensing for qualifying property purchased and placed in service before January 1, 2014. The 50 percent bonus depreciation narrowed any tax differences between eligible assets based on cost recovery provisions, and the 100 percent bonus depreciation eliminated those differences altogether under the provision for allowing a full write-off of asset acquisition costs.	
Credit for Holding New Clean Renewable Energy Bonds	New Clean Renewable Energy Bonds help tax-exempt entities finance capital expenditures for new facilities that produce electricity from renewable sources, including wind energy. Bond holders receive tax credits at 70% of the tax credit interest rate, in lieu of interest payments. New Clean Renewable Energy Bonds may be issued by a public power provider, a cooperative electric company, a governmental body, a clean renewable energy bond lender or a not-for-profit electric utility which has received a loan or loan guarantee under the Rural Electrification Act. The Department of the Treasury publicly solicited applications for an initial volume cap, set by Congress at $800 million, and awarded allocations based on criteria and applications received.	None

Table 9. (Continued)

Initiative name and implementing subagency	Description	Expiration information
	An additional $1.6 billion in New Clean Renewable Energy Bond authorization was provided under the Recovery Act.	
Credit for Holding Qualified Energy Conservation Bonds	Qualified Energy Conservation Bonds provide an opportunity for tax-exempt entities to issue bonds for which bondholders can receive an income tax credit in lieu of interest payments from the issuers of the bonds. Qualified Energy Conservation Bonds can be issued to help finance projects that produce or conserve electricity, including capital expenditures incurred for rural development involving the production of electricity from renewable energy resources, as well as qualified facilities, including wind facilities, under Internal Revenue Code Section 45(d). Qualified Energy Conservation Bonds can be used for a broad array of purposes, including expenditures for certain research facilities, grants, and demonstration projects. Similar to the New Clean Renewable Energy Bonds, the tax credit rate for Qualified Energy Conservation Bonds is 70 percent of the tax credit interest rate. Congress established an initial volume cap of $800 million in Qualified Energy Conservation Bonds that could be issued.	None

Initiative name and implementing subagency	Description	Expiration information
	The cap was raised to $3.2 billion under the Recovery Act. IRS allocated the authority to issue the bonds to U.S. states and territories according to a population-based formula, and the states and territories further allocated the bond issuance authority for individual projects.	
Credit for Residential Energy Efficient Property	The Internal Revenue Code provides an income tax credit to homeowners of up to 30 percent of the costs of eligible energy efficiency equipment, which creates an incentive for homeowners to make such improvements. To be eligible for the credit, the energy efficiency equipment installed by the homeowner must meet certain standards; small wind turbines, solar electric and hot water systems, and geothermal heat pumps are included among the eligible types of equipment. Also, the equipment must be placed in service by December 31, 2016.	Property must be placed in serviceby December 31, 2016.
Direct Payment in Lieu of a Credit for Holding New Clean Renewable Energy Bonds	Through the Hiring Incentives to Restore Employment Act of 2010, Congress permitted issuers of New Clean Renewable Energy Bonds the option to receive a direct payment equivalent to and in lieu of the amount of the tax credit that would otherwise go to the bondholder. This option applied to New Clean Renewable Energy Bonds issued after March 18, 2010.	None

Table 9. (Continued)

Initiative name and implementing subagency	Description	Expiration information
	In cases where bond issuers elect to receive a direct payment, this option helps tax-exempt entities finance projects that produce electricity from renewable sources, including wind energy, because it provides an incentive for investors to purchase the bonds since the investors' returns would not depend upon having sufficient tax liability to utilize a tax credit.	
Direct Payment in Lieu of a Credit for Holding Qualified Energy Conservation Bonds	Through the Hiring Incentives to Restore Employment Act of 2010, Congress permitted issuers of Qualified Energy Conservation Bonds the option to receive a direct payment equivalent to and in lieu of the amount of the tax credit that would otherwise go to the bondholder. This option applied to Qualified Energy Conservation Bonds issued after March 18, 2010. In cases where bond issuers elect to receive a direct payment, this option helps tax-exempt entities finance projects that produce or conserve electricity, including electricity produced from renewable sources such as wind energy, because it provides an incentive for investors to purchase the bonds since the investors' returns would not depend upon having sufficient tax liability to utilize a tax credit.	None

Initiative name and implementing subagency	Description	Expiration information
Energy Investment Credit	The Internal Revenue Code provides an income tax credit based on a percentage of either the cost or fair market value in new equipment that produces electricity and/or heat from renewable energy sources, including wind energy. This credit provides an incentive for businesses to make such investments. Wind, solar, geothermal, and certain other investments qualify for a credit of 30 percent of the investment. Also, the credit is limited based on the date the equipment is placed in service, the production capacity of the equipment, or other factors. Facilities qualifying for the investment tax credit could instead claim a cash grant under Section 1603 of the Recovery Act for facilities placed in service in 2009, 2010, or 2011. For wind facilities placed in service after 2011, taxpayers could claim the investment credit or Section 1603 cash grant if construction of the facility began in 2009, 2010, or 2011, and the property is placed in service before 2013.	Wind projects must begin construction before January 1, 2014, in order to claim the credit.
Energy Production Credit	The Internal Revenue Code provides an income tax credit to businesses and agricultural cooperatives based on the amount of energy they produce at qualified facilities.	Wind projects must begin construction before January 1, 2014, to be eligible.

Table 9. (Continued)

Initiative name and implementing subagency	Description	Expiration information
	The credit provides an incentive for electricity production from a variety of renewable energy sources; although, in calendar year 2011, it was primarily taken by wind energy producers. The amount of the credit varies somewhat depending upon the source and is adjusted annually for inflation. Among other limitations, a taxpayer may generally claim the credit only during the 10-year period commencing with the date the production facility is placed in service, and the credit phases out as the market price of electricity exceeds certain levels. For example, for calendar year 2011, the credit amount was 2.2 cents per kilowatt hour for certain resources (e.g., wind, geothermal, and certain biomass electricity production). For wind facilities placed in service after 2011, taxpayers could claim the cash grant in lieu of the production credit if construction of the facility began in 2009, 2010, or 2011, and the property is placed in service before 2013, or taxpayers could claim the investment credit for facilities placed in service in 2012.	

Initiative name and implementing subagency	Description	Expiration information
Qualifying Advanced Energy Project Credit	The Internal Revenue Code provided an income tax credit of up to 30 percent of the costs to establish, reequip, or expand U.S. domestic manufacturing facilities, which created an incentive to invest in advanced energy equipment manufacturing facilities. Qualified facilities included those that produce equipment designed to produce energy from wind, the sun, geothermal, other renewable resources, as well as other types of energy production and conservation equipment. This credit was established under the Recovery Act, with an initial cap of $2.3 billion in credit allocations. All available amounts were allocated in 2010 but, if some of those credits were forfeited, it may be possible to reallocate them through 2013. A taxpayer claiming this credit was not allowed to also claim the energy investment tax credit.	The application period for the first round of credit certification ended on December 31, 2010. All available amounts were allocated in 2010. Property certified in the first round must be placed in service within 3 years of certification (generally in 2014). If some credits which were allocatedare forfeited, it may be possible to reallocate them through 2013.
Office of Domestic Finance		
Payments for Specific Energy Property in Lieu of Tax Credits	Section 1603 of the Recovery Act established a cash grant program to provide payments to eligible applicants who place specified energy property in service for use in a trade or business. Applicants could take the payment in lieu of either an energy production or investment credit.	Applications must have been submitted before October 1, 2012.

Table 9. (Continued)

Initiative name and implementing subagency	Description	Expiration information
	These payments provide an incentive for investment in property for electricity and heat production, particularly for applicants without sufficient tax liability to utilize a tax credit. Initially, the program provided cash grants for renewable energy projects and production placed in service as of December 31, 2010, or which began construction in 2009 or 2010. However, the program was extended for 1 year as part of the Tax Relief, Unemployment Insurance Reauthorization, and Job Creation Act of 2010. For projects placed in service as of the end of 2011, wind energy projects received the majority of funding, while solar energy project applicants submitted the largest number of funding requests.	

Source: GAO analysis of agency-provided data.

Table 10. Environmental Protection Agency (EPA) Wind-Related Initiatives

Initiative name and implementing subagency	Description	Expiration information
Office of Air and Radiation		
Green Power Partnership	Through this initiative, EPA supports renewable energy development by increasing community and organizational demand for renewable electricity.	None

Initiative name and implementing subagency	Description	Expiration information
	EPA educates prospective partners about green power and offers expert advice, technical support, tools, and resources to encourage organizations to use green power. The program encourages its over 1,000 partnering organizations to use renewable electricity from any and all eligible sources, including wind.	
Office of Research and Development		
People, Prosperity, and the Planet Award Program	Under this initiative, EPA provides competitive grant funding for college student research on sustainable design. Selected teams are awarded $15,000 to develop an idea and demonstrate it at the annual National Sustainable Design Expo in Washington, D.C., where they compete for a $90,000 grant to take their design to market. Research proposals have included projects related to several renewable energy sources, including wind energy.	None
Office of Solid Waste and Emergency Response		
RE-Powering America's Land	Under this initiative, EPA encourages renewable energy development on current and formerly contaminated land and mine sites. EPA has screened over 11,000 sites to determine their potential for renewable energy resources, including wind energy, and provides this information in a Google Earth tool. EPA also provides technical assistance and information on environmental issues related to land contamination and liability for communities, developers, industry, state and local governments, and others interested in reusing these sites for renewable energy development.	None

Source: GAO analysis of agency-provided data.

Table 11. Federal Energy Regulatory Commission Wind-Related Initiatives

Initiative name and implementing subagency	Description	Expiration information
Integration of Variable Energy Resources - Market and Regulatory Reforms to Remove Unduly Discriminatory Practices and Ensure Just and Reasonable Rates	This program implements market reforms that allow all resources, including wind energy, to compete on a level playing field in jurisdictional markets. These efforts include amendments to market rules and the implementation of operational tools that support the reliable integration of renewable resources.	None

Source: GAO analysis of agency-provided data.

Table 12. National Science Foundation (NSF) Wind-Related Initiatives

Initiative name and implementing subagency	Description	Expiration information
Directorate for Engineering		
Energy for Sustainability Program	The program funds fundamental academic research and education activities through grants to enable innovative processes for the sustainable production of electricity and transportation fuels. While the program emphasizes two primary fuel sources—biofuels/bioenergy and photovoltaic solar energy—it also supports research in wind and wave energy, sustainable energy technology assessment, and fuel cells.	None
Energy, Power, and Adaptive Systems	The program provides grants for academic researchers to design and study intelligent and adaptive engineering networks with an emphasis on electric power electronics, networks, and grids. This focus includes generation, distribution, transmission and integration of renewable energy systems and alternative energy technologies, including wind, solar cells, ocean waves, and hydropower. The program also supports laboratory and curriculum development to integrate research and education.	None
Directorate for Engineering - Directorate for Mathematical and Physical Sciences		
Emerging Frontiers in Research and Innovation	The initiative funds grants for interdisciplinary engineering research with the potential to create a significant impact or meet national needs.	None

Initiative name and implementing subagency	Description	Expiration information
	Under the program, NSF has coordinated with DOE to call for proposals that help address the need for wind and solar energy storage.	

Source: GAO analysis of agency-provided data.

Table 13. Small Business Administration (SBA) Wind-Related Initiatives

Initiative name and implementing subagency	Description	Expiration information
Office of Capital Access		
Certified Development Company / Section 504 Loans	These loan guarantees supplement SBA loans to businesses investing in energy efficient equipment, renewable energy sources—such as wind energy—or equipment, or green buildings. SBA allows green building and energy efficient activities to qualify as broader public policy goals, allowing for larger loans and lower job creation goals for eligible companies.	None
Office of Investment		
Energy Saving Debenture	This initiative provides a financing tool for Small Business Investment Companies licensed after 2008 to invest in small businesses with products or services that are directed toward energy saving activities, including renewable energy. Energy Saving Debentures provide a financing option for Small Business Investment Companies to make energy saving qualified investments in these small businesses.	None

Source: GAO analysis of agency-provided data.

APPENDIX III: ESTIMATED REVENUE LOSSES FOR TREASURY'S WIND-RELATED TAX EXPENDITURES

Table 14 below reflects the fiscal year 2011 revenue loss estimates for Treasury's wind-related tax expenditures—both their total estimated revenue losses and their estimated revenue losses for activities specifically related to wind.[1]

Table 14. Estimated Revenue Losses for Treasury's Wind-Related Tax Expenditures in Fiscal Year 2011

Initiative name	Estimated revenue losses	Estimated revenue losses foractivities specifically related to wind[a]
Accelerated Depreciation Recovery Periods for Specific Energy Property	Up to $350 million[b]	Estimates are not available
Credit for Holding New Clean Renewable Energy Bonds	$70 million	Estimates are not available
Credit for Holding Qualified Energy Conservation Bonds	$10 million	Estimates are not available
Credit for Residential Energy Efficient Property	$840 million	Estimates are not available
Direct Payment in Lieu of a Credit for Holding New Clean Renewable Energy Bonds	$20 million	Estimates are not available
Direct Payment in Lieu of a Credit for Holding Qualified Energy Conservation Bonds	$30 million	Estimates are not available
Energy Investment Credit (ITC)[c]	$700 million	Less than $50 million for small windproperties—properties using wind turbines of100 kilowatts or less[b]
Energy Production Credit (PTC)[d]	$1,560 million	$1,100 million[b]
Qualifying Advanced Energy Project Credit	$430 million	Estimates are not available

Sources: GAO analysis of Office of Management and Budget and Joint Committee on Taxation documentation.

Note: Revenue losses and outlay effects reflect Treasury estimates from the President's Fiscal Year 2013 budget unless otherwise specified. Revenue loss estimates do not incorporate any behavioral responses and thus do not reflect the exact amount of revenue that would be gained if a specific tax expenditure were repealed. In addition, while sufficiently reliable as a gauge of general magnitude, summing individual tax expenditures' revenue loss estimates does not take into account interactions between individual provisions.

[a] The Office of Management and Budget—which uses revenue loss estimates developed by Treasury—does not publish energy source-specific revenue loss estimates.

[b] The revenue loss estimate is reported by the Joint Committee on Taxation only.

[c] Firms can take a cash grant under Treasury's Payments for Specific Energy Property in Lieu of Tax Credits initiative (Section 1603 program) instead of the ITC for facilities placed in service in 2009, 2010, and 2011, or if their construction began in 2009, 2010, or 2011.

[d] Firms can take a cash grant under Treasury's Section 1603 program instead of the PTC for facilities placed in service in 2009, 2010, or 2011, or if their construction began in 2009, 2010, or 2011.

Appendix IV: GAO's Questionnaire for Federal Agencies with Wind-Related Initiatives

QUESTIONS ABOUT FEDERAL WIND INITIATIVES – *[PRE-POPULATED INFORMATION]*

GAO
Accountability * Integrity * Reliability

United States Government Accountability Office

QUESTIONS ABOUT FEDERAL WIND-RELATED INITIATIVE:
[PRE-POPULATED INFORMATION]

Introduction
The United States Government Accountability Office (GAO), an independent, legislative-branch agency, is examining how the federal government promotes research and development, commercialization, and deployment of wind energy through wind-related initiatives (GAO job code 361379). We define an initiative as a group of programmatic or mission-area activities serving similar purposes or functions. We are not considering individual projects within an initiative—such as a specific grant award—to be initiatives in themselves. Wind-related initiatives are those that could or do promote wind energy, either exclusively or as part of a broader initiative. GAO is beginning this work in response to a congressional mandate in the Statutory Pay-As-You-Go Act of 2010, Public Law No. 111-139, Section 21. The objectives for this study are to examine: (1) the key characteristics of wind energy initiatives supported by the federal government; (2) the extent to which, if any, there is fragmentation, overlap, and duplication of federal wind energy initiatives; and (3) the extent to which agencies review their portfolios of wind energy initiatives and coordinate similar efforts internally and with other agencies.

In 2011, we asked your agency to provide us with information and data on the renewable energy initiatives that were ongoing at your agency in fiscal year 2010 (GAO job code 361185). Using that information, and information provided to us in the prior questionnaire distributed for this engagement on wind-related initiatives, we have identified initiatives that your agency indicated promoted wind energy through research and development, commercialization, or deployment, and were active in fiscal year 2011. Through this supplemental questionnaire, we are asking you to provide additional information about the ***[PRE-POPULATED INFORMATION]*** initiative. After we receive your response, we may follow up to discuss and clarify any outstanding questions about the initiative relative to the objectives stated above.

Please complete all four sections and return it to one of the individuals mentioned below by **May 1, 2012**. When returning the questionnaire, please attach any relevant supporting documentation to your e-mail.

If you have any questions or comments about this questionnaire, please call or e-mail Miles Ingram at (617) 788-0549 (ingramm@gao.gov) or Keya Chateauneuf at (617) 788-0583 (chateauneufk@gao.gov). Thank you very much for your assistance.

Page 1 of 11

QUESTIONS ABOUT FEDERAL WIND INITIATIVES – *[PRE-POPULATED INFORMATION]*

Instructions

This questionnaire can be completed using MS-Word and returned via e-mail to ingramm@gao.gov or chateauneufk@gao.gov. Please complete this questionnaire and return it by **May 1, 2012**.

1) Use your mouse to navigate by clicking on the check box ☐ or answer box you wish to answer.

2) To select a check box, click on the center of the box, and an 'X' will appear.

3) To deselect a check box response, click on the center of the box, and the 'X' will disappear.

4) To answer a question that requires a comment, click on the answer box and begin typing. The box will expand to accommodate your answer.

QUESTIONS ABOUT FEDERAL WIND INITIATIVES – *[PRE-POPULATED INFORMATION]*

SECTION I: GENERAL INFORMATION ABOUT INITIATIVE

We define wind-related initiatives as those that could or do promote or enable wind energy development—either exclusively or as part of a broader initiative. The initiative itself may support solely, or in part, wind energy *research and development*—including efforts ranging from basic research defining scientific concepts to efforts to apply and demonstrate new and improved technologies. The initiative may also support solely, or in part, wind energy *commercialization*—such as, efforts to bridge the gap between research and development activities and the marketplace through transitioning technologies to commercial applications. In addition, the initiative may support wind energy *deployment*–efforts to facilitate or achieve widespread use of technologies, such as through their construction or operation. Under our definition of deployment, we are also including regulatory initiatives that focus on streamlining regulatory processes, providing regulatory access to resources, or fast-tracking permitting of certain projects.

Your agency provided the following information about the initiative to a previous GAO job (job code 361185).

Initiative Name: *[PRE-POPULATED INFORMATION]*

Implementing Sub-Agency: *[PRE-POPULATED INFORMATION]*

Implementing Office: *[PRE-POPULATED INFORMATION]*

Initiative Description: *[PRE-POPULATED INFORMATION]*

Recipient Type: *[PRE-POPULATED INFORMATION]*

Expiration/sunset date: *[PRE-POPULATED INFORMATION]*

1. **Is the information provided above about the initiative correct and complete for fiscal year 2011?**
   Yes.......................... ☐
   No........................... ☐

If you answered No, what information is incorrect or incomplete, and what is the correct or complete information? Please use the space below.

QUESTIONS ABOUT FEDERAL WIND INITIATIVES – *[PRE-POPULATED INFORMATION]*

2. When did wind energy first receive or become eligible for support under this initiative?

Please check one box in each row.	Fiscal year 2011 ▼	Fiscal year 2010 ▼	Fiscal year 2009 ▼	Fiscal year 2008 or before ▼	Don't know/Not able to determine ▼
a) Wind energy first received support..........	☐	☐	☐	☐	☐
b) Wind energy was first eligible for support..	☐	☐	☐	☐	☐

3. On which of the following areas does your initiative focus?

Please check Yes or No for each category.	Yes ▼	No ▼
Utility-scale onshore wind	☐	☐
Distributed (small-scale) onshore wind	☐	☐
Offshore wind	☐	☐
Transmission	☐	☐
Grid integration	☐	☐
Siting and permitting	☐	☐
Other? *Please specify in the space below.*	☐	☐

Page 4 of 11

QUESTIONS ABOUT FEDERAL WIND INITIATIVES – *[PRE-POPULATED INFORMATION]*

4. We understand that each agency defines technology-related activities in a unique manner; however, in order to compare activities across agencies, we have developed categories that can be applied broadly and have listed them below. Please determine which categories best fit the wind-related activities under your initiative.

Which of the following technology-related categories apply to this initiative's wind-related activities? *Please check Yes or No for each category.*	Yes ▼	No ▼
a. Basic research? .... *Basic research includes efforts that explore and define scientific or engineering concepts, or is conducted to investigate the nature of a subject without targeting any specific technology.*	☐	☐
b. Applied research? .... *Applied research includes efforts that develop new scientific or engineering knowledge to create new and improved technologies.*	☐	☐
c. Demonstration activities? .... *Demonstration activities include efforts that operate new or improved technologies to collect information on their performance and assess their readiness for widespread use.*	☐	☐
d. Commercialization? .... *Commercialization includes efforts that bridge the gap between research and demonstration activities, and venture capital funding and marketing activities, through transitioning technologies to commercial applications.*	☐	☐
e. Deployment? .... *Deployment includes efforts that facilitate or achieve widespread use of technologies in the commercial market through their construction, operation or use. We are also including initiatives that focus on streamlining regulatory processes, providing regulatory access to resources, or fast-tracking permitting of certain projects.*	☐	☐
f. Other? *Please specify in the space below*	☐	☐

QUESTIONS ABOUT FEDERAL WIND INITIATIVES – *[PRE-POPULATED INFORMATION]*

5. Please describe the overall goals of your initiative in the space below. Briefly describe the source from which the goals come, such as program-level documents, agency-wide documents, or government-wide directives from the Office of Management and Budget or Congress.

6. If the above goals are not specific to wind energy, does your initiative have goals that <u>are specific to wind energy</u>?

N/A - wind-specific goals provided in Question 5... ☐
Yes.. ☐
No... ☐

If you answered Yes, please briefly describe the goals for wind energy in the space below. Please provide a link(s) to the applicable website(s) in the space below or some other reference for documents such as strategic plans, annual reports, program plans, grant solicitations, or requests for proposals. Please cite the relevant portion of the document (e.g. document name and page number) if applicable.

7. Has your initiative created performance measures to track progress toward the <u>wind energy</u> goals?

Yes.. ☐
No... ☐
N/A.. ☐

If you answered Yes, please describe the performance measures in the space below.

8. Has anyone at <u>your agency</u> (including any sub-agencies or offices) or any third party assessed this initiative to determine the extent to which it has met any of its goals or targets?

Yes .... ☐
No.......................... ☐
Don't know .... ☐

If yes, please describe the assessment(s), including whether or not a report was generated, in the space below. If you have a report or other documentation, please provide a weblink or attach a copy.

Page 6 of 11

QUESTIONS ABOUT FEDERAL WIND INITIATIVES – *[PRE-POPULATED INFORMATION]*

SECTION II: FUNDING FOR INITIATIVE

We are interested in how much money goes toward the wind-related activities of your initiative. We would like to be able to report obligations data across all federal initiatives for fiscal year 2011. If necessary, please consult with staff in your agency's budget office to answer these funding-related questions.

For questions in this section, please use the following definitions:

- *An **obligation** is a definite commitment that creates a legal liability of the government for the payment of goods and services ordered or received, or a legal duty on the part of the United States that could mature into a legal liability. Payment may be made immediately or in the future. An agency incurs an obligation, for example, when it places an order, signs a contract, awards a grant, purchases a service, or takes other actions that require the government to make payments to the public or from one government account to another.*
- ***Wind-related activities** are those that could or do promote or enable the development of wind energy. For initiatives with a focus on broader electricity issues, such as transmission or grid integration, we would consider activities to be wind-related if they promote or enable the development of wind energy, even if this is not an explicitly stated goal of the activity. Similarly, for initiatives such as basic science research programs that do not directly promote or enable development of wind-specific technologies, we would consider activities to be wind-related if they could promote or enable the future development of wind energy.*

Note: We may request copies of supporting documentation for the numbers you provide below.

9. **For fiscal year 2011, please provide actual or estimated obligations data for <u>[PRE-POPULATED INFORMATION]'s wind-related activities</u>.** We understand that some agencies may not be able to provide actual or estimated obligations for certain initiatives.

Can you provide **obligations** data for this initiative's **wind related activities for FY11?**	**If yes, what were the obligations** for this initiative's **wind related activities for FY11?**	Is this amount an **estimate?**
☐ Yes →	$____ →	☐ Yes
☐ No		☐ No

If you were unable to provide actual or estimated obligations, please explain why in the space below. If you are able to provide funding data other than obligations for your initiative (e.g. outlays, loan guarantee amounts, borrowing authority, etc), please explain here and provide what is available for fiscal year 2011.

QUESTIONS ABOUT FEDERAL WIND INITIATIVES – *[PRE-POPULATED INFORMATION]*

10. Government auditing standards require that we assess the reliability of data used in our products. Therefore, **for the data you reported in the previous question,** we would like to obtain answers to the following questions about the data and/or the information system that produces them (e.g. databases or other electronic records).

   a. **If the data provided in question 9 are an estimate, please explain how the estimate was determined and identify what limitations, if any, are associated with this estimate.**

   b. **What is the source of the data provided in question 9** (e.g. name of database or other electronic records system)?

   c. **Please describe any data system controls, edit checks, or internal auditing procedures in place that ensure the accuracy and completeness of the data. What procedures, if any, are in place to review and certify the reliability of this data? Have there been any reviews of the quality of the data** (e.g. inspector general or internal audit reports, internal reviews and studies, or contractor or consultant studies)? If so, please provide copies or web links.

   d. If you needed to consult with someone more involved with relevant data system(s) to answer this question, **please identify that individual and provide contact information so that we may ask follow-up questions, if necessary.**

11. As part of our review, we'd like to provide a few examples of projects or program activities that have been funded under this initiative. **Please briefly describe two examples of specific wind-related projects or program activities that were active under this initiative in fiscal year 2011 in the space below. (Please limit your response to 800 characters.)**

   a. Example #1

   b. Example #2

12. **Considering the range of projects or activities supported and the resources committed to these projects or activities in fiscal year 2011, which of the following best describes this initiative?**

Solely supported wind-related activities.......................................☐
Primarily supported wind-related activities....................................☐
Supported wind-related and non wind-related activities equally.....☐
Primarily supported activities that are <u>not</u> wind-related...............☐
Don't know/Not able to determine..☐

If you don't know or are not able to determine, please explain:

Page 8 of 11

SECTION III: COORDINATION

13. Do you or staff involved with the [PRE-POPULATED INFORMATION] initiative *formally* coordinate with other wind-related initiatives <u>within your agency</u> or <u>with other federal agencies</u>? An example of formal coordination might be participation in a working group or having a memorandum of understanding or interagency agreement.

Yes..........................☐
No...........................☐

If you answered Yes to the above question, please provide up to five examples in the table below of how you coordinate within your agency or with other federal agencies.

Name/Title of coordination activities	Description of coordination activities	Does this coordination activity involve agencies <u>outside of your agency</u>? Yes ▼	No ▼
1.		☐	☐
2.		☐	☐
3.		☐	☐
4.		☐	☐
5.		☐	☐

If you have additional information you would like to provide about your coordination activities, please describe in the space below:

14. Do you or staff involved with this initiative *formally* coordinate your initiative's activities with __state governments__? An example of formal coordination might be participation in a working group or having a memorandum of understanding or interagency agreement.

Yes ☐
No ☐

If you answered Yes to the above question, please provide up to five examples in the table below of how you coordinate with state governments.

Name/Title of coordination activities	Description of coordination activities
1.	
2.	
3.	
4.	
5.	

If you have additional information you would like to provide about your coordination activities, please describe in the space below:

15. What steps, if any, does __your agency__ (or your sub-agency or office) take to assess potential overlap or duplication of its initiatives' activities with other federal efforts?

16. When reviewing requests for funding, such as grant applications or bids for contracts, do you take steps to determine whether or not the requesters have received other sources of federal funding for the project?

Yes ☐
No ☐

If you answered Yes, please describe in the space below what information you review and how you determine whether or not they have received other sources of federal funding for the project.

SECTION IV: OTHER INFORMATION

17. Are there any additional data, nuances, further sources of information, or comments not covered previously that would help us further understand and report on how this initiative is being implemented? If so, please describe and/or provide link(s) to the relevant website(s) in the space below. *If you do not have additional information to provide, it is appropriate to leave this question blank.* If applicable, please provide electronic copies of the relevant documents (e.g., strategic planning documents, performance metrics, requests for proposals, regulatory provisions, etc.) when returning this questionnaire. Alternatively, you can suggest that further discussion about this initiative be conducted through follow-up conversations with GAO.

18. Please provide the contact information for a representative from your agency for any follow-up questions we may have about this questionnaire.

Name:

Title:

Agency:

Office/Division:

Email:

Phone:

Please remember to attach any relevant supporting documentation when returning this questionnaire to either or ingramm@gao.gov or chateauneufk@gao.gov.

Thank you for your time!

End Notes

[1] Although this report focuses on federal support for wind energy, the federal government has provided substantial support for other renewable and fossil energy sources as well as nuclear energy. We have recently examined support for some of these energy sources and have other ongoing work in the area that will address federal subsidies for energy. See GAO, *Solar Energy: Federal Initiatives Overlap but Take Measures to Avoid Duplication*, GAO-12-843 (Washington, D.C.: Aug. 30, 2012); GAO, *Batteries and Energy Storage: Federal Initiatives Supported Similar Technologies and Goals but Had Key Differences*, GAO-12-842 (Washington, D.C.: Aug. 30, 2012); GAO, *Research and Development: DOE Could Enhance the Project Selection Process for Government Oil and Natural Gas Research,* GAO-09-186 (Washington, D.C.: Dec. 29, 2008); GAO, *Oil and Gas Royalties: The Federal System for Collecting Oil and Gas Revenues Needs Comprehensive Reassessment*, GAO-08-691 (Washington, D.C.: Sept. 3, 2008).

[2] Pub. L. No. 111-5, 123 Stat. 115 (2009).

[3] The subsidies for wind energy reported in fiscal years 2007 and 2010 may not be directly comparable due to differences in available incentives during the 2 years. Specifically, beginning in 2009, Section 1603 of the Recovery Act, which created Treasury's Payments for Specific Energy Property in Lieu of Tax Credits initiative (Section 1603 program), allowed recipients to choose a direct grant in lieu of the Energy Production Credit (also known as the Production Tax Credit or PTC). The Section 1603 program, which was unavailable in 2007, and was popular with investors in fiscal year 2010, front-loads the government's support for projects in the year that the grant is awarded, as opposed to the PTC, which spreads out support over 10 years. Thus, while both incentives may provide the same overall level of support to recipients, the amount of support provided in a given year will differ based on which incentive is taken.

[4] For purposes of this report, "tax subsidies" refer to the benefits provided to taxpayers who take advantage of tax expenditures, and thus pay lower taxes than they would otherwise have had to pay. Tax expenditures are provisions of federal tax laws that (1) allow a special exclusion, exemption, or deduction from gross income or (2) provide a special credit, preferential tax rate, or deferral of tax liability. Tax expenditures result in revenue losses for the federal government, which forgoes some of the tax revenues that it would have otherwise collected. See GAO, *Government Performance and Accountability: Tax Expenditures Represent a Substantial Federal Commitment and Need to Be Reexamined*, GAO-05-690 (Washington, D.C.: Sept. 23, 2005).

[5] GAO, *Renewable Energy: Federal Agencies Implement Hundreds of Initiatives*, GAO-12-260 (Washington, D.C.: Feb. 27, 2012).

[6] For the purposes of this report, wind issues covered by wind-related initiatives include utility-scale land-based wind, distributed land-based wind (also known as small wind), offshore wind, transmission, grid integration, and siting and permitting.

[7] For the purposes of this report, technology advancement activities include basic research, applied research, demonstration, commercialization, and deployment.

[8] GAO, *Managing for Results: Using the Results Act to Address Mission Fragmentation and Program Overlap*, GAO/AIMD-97-146 (Washington, D.C.: Aug. 29, 1997). For more information on fragmentation, overlap, and duplication in federal programs see GAO, *Opportunities to Reduce Potential Duplication in Government Programs, Save Tax Dollars, and Enhance Revenue*, GAO-11-318SP (Washington, D.C.: Mar. 1, 2011).

[9] For purposes of this report, we defined a wind-related initiative as a program or group of agency activities serving a similar purpose or function that promoted wind energy technologies through a specific emphasis or focus, even if wind energy was only one part of a broader effort.

[10] A revenue loss is the amount of revenue that the government forgoes as the result of each special provision in the tax code. Revenue loss is estimated for each tax expenditure separately by comparing the revenue raised under current law with the revenue that would have been raised if the provision did not exist (assuming that taxpayer behavior and all other tax and spending provisions remain constant). Revenue loss estimates do not incorporate any behavioral responses and thus do not necessarily represent the exact amount of revenue that would be gained if a specific tax expenditure were repealed.

[11] In the United States, all utility-scale wind farms currently in place are built on land. Offshore wind turbines are utility-scale turbines that capture wind resources over bodies of water and convert this wind into electricity. Although there are currently no offshore wind turbines in United States, there are federal, regional, and state-level efforts intended to facilitate their development.

[12] We developed definitions that could be applied broadly to make comparisons across agencies and that covered the full spectrum of advancement activities. Federal agencies use various definitions and categories for describing the stages of technology advancement. For example, DOE uses technology readiness level categories and definitions to measure and communicate technology readiness for first-of-a-kind technology applications. In addition, the Office of Management and Budget provides federal definitions for the terms basic research, applied research, and development in Office of Management and Budget Circular A–11, Section 84—Character Classification (Schedule C) but does not provide definitions for demonstration activities, commercialization, or deployment. During pretests of our questionnaire, agency officials were able to fit their initiatives' activities within the categories that we defined.

[13] Congressional Budget Office, *Federal Financial Support for the Development and Production of Fuels and Energy Technologies* (Washington, D.C.: March 2012).

[14] DOE, *2011 Wind Technologies Market Report* (Washington, D.C.: 2012).

[15] DOE, Energy Information Administration, *Direct Federal Financial Interventions and Subsidies in Energy in Fiscal Year 2010* (Washington, D.C.: 2011).

[16] Our review focused on initiatives that promoted the research and development, commercialization, or deployment of wind energy technologies in fiscal year 2011. In addition, we excluded certain agencies—such as the Departments of Defense, Homeland Security, and State—which have initiatives generally focused on development of wind energy and other technologies for use in a military, border security, or international aid setting, rather than for use in the domestic commercial energy market. We also excluded the Departments of Transportation and Housing and Urban Development because their initiatives focused on commercialization and deployment of renewable energy projects on buildings, projects that typically use solar energy.

[17] The federal obligations and tax subsidies for fiscal year 2011 presented here cannot be compared to the Energy Information Administration's (EIA) estimate of $5 billion in total federal subsidies for wind in fiscal year 2010 because of differences in the time period and methods used in calculating these numbers. For instance, EIA measured subsidies on the basis of the cost of the programs to the federal budget as provided in budget documents from the Office of Management and Budget and the Joint Committee on Taxation. Our data

and analysis of obligations and tax subsidies was based on agency questionnaire responses and interviews with officials.

[18] In addition, agency officials for 18 percent of initiatives were not able to determine to what extent their initiatives supported wind energy relative to other sources of renewable energy or other activities. In some cases, officials were unable to make these determinations because of data limitations. For example, several initiatives did not track program data separately for each energy source.

[19] On October 1, 2011, BOEMRE reorganized into two independent entities: the Bureau of Ocean Energy Management and the Bureau of Safety and Environmental Enforcement. The Bureau of Ocean Energy Management is responsible for managing development of the nation's offshore resources in an environmentally and economically responsible way, and its activities include oversight of leasing, environmental studies, and economic analysis. The Bureau of Safety and Environmental Enforcement is responsible for enforcing safety and environmental regulations. However, given the time frame of our work, we refer to BOEMRE in this report.

[20] Carbon capture and storage is a process of separating carbon dioxide from other gases produced in fuel combustion and other industrial processes, transporting the carbon dioxide via pipeline to an underground storage location, and injecting and storing it longterm in underground geologic formations.

[21] Nearly all of these obligations were incurred by initiatives established under the Recovery Act. However, the Recovery Act initiatives with the most obligations for activities specifically related to wind—Treasury's Section 1603 program and DOE's Title XVII Section 1705 Loan Guarantee Program—were alternatives to initiatives that were established prior to the Recovery Act—the PTC and Energy Investment Credit (also known as the Investment Tax Credit, or ITC), as well as the Title XVII Section 1703 Loan Guarantee Program. Some projects the Recovery Act initiatives supported in fiscal year 2011 could have instead received support from the ongoing initiatives that existed prior to the act.

[22] As previously noted, most wind-related initiatives supported a range of energy sources in addition to wind. We asked agency officials to provide actual or estimated obligations data for their initiative's activities that were specifically related to wind. In addition, unlike spending data presented in the President's budget—which are estimated for fiscal years that are not completed at the time the budget is prepared—the estimated obligations data presented here are for completed fiscal years.

[23] In addition to the 21 initiatives for which no obligations data were reported, 9 of the initiatives (about 11 percent) for which officials did not provide obligations data were Treasury's wind-related tax expenditures, discussed below. Note that the percentages add to 101 due to rounding.

[24] In addition to the PTC, which provided an estimated $1.1 billion in wind-specific tax subsidies in 2011, the ITC provided less than $50 million in subsidies in 2011 for small wind properties—properties using wind turbines of 100 kilowatts or less—according to the Joint Committee on Taxation. However, this estimate does not include ITC support for larger wind properties. See Joint Committee on Taxation, *Estimates of Federal Tax Expenditures for Fiscal Years 2011-2015,* JCS-1-12 (Washington, D.C.: Jan. 17, 2012). Also, Treasury has developed wind-specific estimates of subsidies from the Credit for Residential Energy Efficient Property, but the most recent year for which data are available is 2009. Treasury officials said these estimates are typically reported 2 years after the subsidies are provided.

[25] In addition to the 63 initiatives supporting more than one wind issue, responses to our questionnaires for 15 initiatives indicated support for only one wind issue, and responses for

4 other initiatives indicated only an "other" wind issue. The numbers of initiatives supporting one or more than one wind issue does not reflect whether officials responded that initiatives supported "other" wind issues in addition to one of the six main wind issues.

[26] In addition to the 39 initiatives supporting one type of technology advancement activity and the 39 initiatives supporting more than one, responses to our questionnaires for 2 initiatives indicated only an "other" technology advancement activity, and responses for 2 others did not indicate any technology advancement activity. The numbers of initiatives supporting one or more than one technology advancement activity do not reflect whether officials responded that initiatives supported "other" technology advancement activities in addition to one of the five main types of activities.

[27] The amount of the credit varies by energy source. For example, the amount of the credit is currently 1.1 cents per kilowatt hour for energy produced from a subset of qualified energy sources, including hydropower, landfill gas, and others. In addition, the amount of the credit is adjusted annually for inflation. For example, when it was first created under the Energy Policy Act of 1992, the amount of the credit was 1.5 cents per kilowatt hour for energy produced by wind facilities.

[28] The number of DOE initiatives supporting deployment was equal to or greater than the number of its initiatives supporting any other technology advancement activity.

[29] GAO-11-318SP.

[30] Compressed air energy storage involves injecting compressed air into a geological formation such as an underground cavern within a salt dome. To recover the power, the air is released and used to help drive a turbine generator. Compressed air energy storage provides bulk energy storage for stationary power systems.

[31] Gilbert Metcalf, *Investment in Energy Infrastructure and the Tax Code*, From the Selected Works of Gilbert E. Metcalf (2009); Sanya Carley, *State Renewable Energy Electricity Policies: An Empirical Evaluation of Effectiveness*, Energy Policy (2009); Gireesh Shrimali and Joshua Kniefel, *Are Government Policies Effective In Promoting Deployment of Renewable Electricity Resources?* Energy Policy (2011); Haitao Yin and Nicholas Powers, *Do State Renewable Portfolio Standards Promote In-State Renewable Generation?* Energy Policy (2010).

[32] State RPS data are from DSIRE. Of the 37 states, 29 states have legally binding state standards, and 8 states have nonbinding RPS goals, according to DSIRE data. In addition to these states, Florida has a municipal utility that has signed a memorandum of understanding formalizing its commitment to generate at least 7.5 percent of its electric capacity from wind and other renewable energy sources by 2015.

[33] Texas law established a target of 500 megawatts of the total required 5,880 megawatts to come from sources other than wind energy. As of the end of 2011, 10,394 megawatts of wind energy capacity had been installed in Texas according to DOE.

[34] We did not include informal coordination in the scope of this report. Our past work has shown that informal coordination mechanisms may be dependent on relationships among individual officials and that these informal relationships could end if personnel move to different assignments. See GAO, *National Security: Key Challenges and Solutions to Strengthen Interagency Collaboration,* GAO-10-822T (Washington, D.C.: June 9, 2010).

[35] GAO, *The Government Performance and Results Act: 1997 Governmentwide Implementation Will Be Uneven*, GAO/GGD-97-109 (Washington, D.C.: June 2, 1997); GAO, *2012 Annual Report: Opportunities to Reduce Duplication, Overlap and Fragmentation, Achieve Savings, and Enhance Revenue*, GAO-12-342SP (Washington, D.C.: Feb. 28, 2012).

[36] GAO, *Results-Oriented Government: Practices That Can Help Enhance and Sustain Collaboration among Federal Agencies,* GAO-06-15 (Washington, D.C.: Oct. 21, 2005).

[37] See 42 U.S.C. § 16512; 10 C.F.R. §609.9 (d); and Office of Management and Budget Circular A-129 Revised (Nov. 2000), *Policies for Federal Credit Programs and Non-Tax Receivables.*

[38] For instance, according to DOE officials, this type of arrangement can be made on a "pro rata" basis: that is, the percentage of the proceeds from the grant used to repay debt is the same as the percentage of a project's cost financed by debt. For example, if 70 percent of a project's capital costs are financed by debt, then 70 percent of Section 1603 grant proceeds will be used to pay down the debt. The percentage of debt financing varies from project to project.

[39] We focused our analysis of duplication on initiatives at the five lead agencies (DOE, Interior, USDA, Commerce and Treasury) that provided financial support to energy industry recipients for the construction or operation of utility-scale land-based wind facilities. These initiatives represented the vast majority of actual and estimated wind-related obligations and tax subsidies in fiscal year 2011. For more detailed discussion of our methodology for identifying duplication, see app. I.

[40] Applications for Section 1603 grants must have been submitted before October 1, 2012, though Treasury officials noted that they anticipate significant awards to applicants will be made under the Section 1603 program over the next 4 years.

[41] Grant, loan, and tax expenditure initiatives provide capital for wind and other eligible projects—that is, the funds or tax subsidies provided by the government are used to cover the costs of developing and building a project. In contrast, loan guarantees provided through initiatives such as DOE's Section 1703 and Section 1705 programs lower the cost of capital for projects by helping companies obtain affordable financing because the federal government agrees to reimburse the lender for the guaranteed amount if a borrower defaults. Despite this difference between these forms of government support, both types of initiatives provide a financial benefit for developers of wind projects.

[42] Lawrence Berkeley National Laboratory, *Avoiding the Haircut: Potential Ways to Enhance the Value of the USDA's Section 9006 Program* (Berkeley, CA: July 2006).

[43] National Renewable Energy Laboratory and Lawrence Berkeley National Laboratory, *PTC, ITC, or Cash Grant? An Analysis of the Choice Facing Renewable Power Projects in the United States* (Golden, CO and Berkeley, CA: March 2009); Lawrence Berkeley National Laboratory, *Revealing the Hidden Value that the Federal Investment Tax Credit and Treasury Cash Grant Provide To Community Wind Projects* (Berkeley, CA: January 2010); Lawrence Berkeley National Laboratory, *Community Wind: Once Again Pushing the Envelope of Project Finance* (Berkeley, CA: January 2011).

[44] Developers of wind projects can cover the costs of their projects using several sources of capital, including debt and equity. Tax equity is a form of capital provided for projects by investors in exchange for tax benefits—primarily the PTC and accelerated depreciation—and a share of the revenues. Tax equity investors allow developers without sufficient tax liabilities to capture the benefit of tax credits and other tax subsidies. Using tax equity, a developer can convert its federal tax benefits into cash through an investor, at a discount, and use this cash to pay for a portion of project capital and installation costs.

[45] Project developers may not receive the full benefit of tax subsidies such as the PTC and accelerated depreciation, because the subsidies are often shared between the developers and financial or other institutions that invest tax equity in projects. Section 1603 grant payments go to whatever entity places the wind project into service, whether that is the developer or a tax equity investor. The Section 1603 program allows developers without sufficient tax

liabilities to forgo the use of tax equity investors and directly receive the full amount of the government subsidy. However, developers may still partner with such investors who can take advantage of tax benefits such as depreciation, and who may also receive part of the subsidy provided through Section 1603 grants. According to financial professionals with whom we spoke, the rates of return on tax equity for such investors generally range from 7 to 9 percent for typical projects and can range up to twice that amount for riskier projects.

[46] 26 U.S.C. § 45 (b)(3). This provision includes an exception for certain closed-loop biomass facilities. Loan guarantees provided under DOE's Section 1703 and Section 1705 programs are not treated as subsidized energy financing under this provision. The amount of the credit is not reduced by any state tax credits received in connection with a project. See Rev. Rul. 2006-9, 2006-9 I.R.B. 519.

[47] Capacity factors represent the electricity generated by a facility as a percentage of that facility's maximum capacity to generate electricity.

[48] According to officials, Treasury determines on a case-by-case basis whether grants under REAP or other federal grant programs must be taxed as income. For instance, officials said grants are generally not taxed if they meet the criteria of the "general welfare exclusion." This exclusion applies to payments made under social benefit programs for the promotion of general welfare based upon the recipient's need. Such payments include federally provided mortgage assistance payments to low-income homeowners, or certain government payments to victims of a natural disaster.

[49] The Recovery Act eliminated the provision reducing the value of the ITC and Section 1603 program grants for government-subsidized, low-interest loans, though such a provision remains for the PTC.

[50] The Recovery Act allows a partial exception to this provision for projects receiving Section 1603 grants along with accelerated depreciation. In this case, a project's depreciable basis is reduced by half of the grant's value—15 percent of project costs— rather than the full 30 percent of project costs that Section 1603 grants are worth.

[51] DOE officials told us that, under applicable provisions of the appropriations laws, the Office of Management and Budget is required to certify compliance with this restriction. The restriction appears in the fiscal year 2009 Omnibus Appropriations Act (Pub. L. No. 111-8), as well as the Department of Defense and Full-Year Continuing Appropriations Act, 2011, (Pub. L. No. 112-10) and provides that DOE may not provide loan guarantees based on its authority under that act to projects that benefit from other federal grants or support (other than tax benefits, federal leases, insurance, and transmission facilities).

[52] Rev. Rul. 2006-9, 2006-9 I.R.B. 519.

[53] As noted above, project developers may not receive the full benefit of tax subsidies such as the PTC and accelerated depreciation, because the subsidies are often shared between the developers and financial or other institutions that provide tax equity for projects. In addition, regardless of whether the tax subsidies are provided to developers or tax equity investors, the full benefit of these subsidies may not always be achieved because either (1) the taxpayer may have insufficient taxable income to fully use the credits or depreciation deductions over the period of years they are provided, or (2) the taxpayer may be required to pay the Alternative Minimum Tax due to the use of the credits and deductions.

[54] Financial professionals' estimates for accelerated depreciation ranged from 10 percent to 30 percent of projects' capital costs. Estimates in the low end of this range accounted for only the incremental value provided for wind properties by 5-year accelerated depreciation, above the value of straight-line depreciation. Under the straight-line method, annual depreciation expenses are calculated by dividing the purchase price of the asset (less its

estimated salvage or residual value) by the estimated useful life of the asset. Estimates in the high end of the range included the full amount of deductions permitted under accelerated depreciation. Treasury officials noted that these estimates of accelerated depreciation reflect its tax value in absolute terms, but do not reflect whether depreciation allowances granted to wind projects provide a greater benefit than do depreciation allowances for other industries. As noted above, the 50 percent bonus depreciation allowed under current law narrows any tax differences between eligible assets, and the 100 percent bonus depreciation allowed under previous legislation eliminated those differences altogether by allowing a full write-off of asset acquisition costs.

[55] Booz & Company Inc.—a global management consulting firm for businesses, governments, and other organizations—developed these estimates using a proprietary economic model that it uses to advise its clients in the energy industry. We did not independently verify the model's calculations.

[56] We focused our analysis of agencies' allocation of support to projects and assessment of applicant need on the 10 initiatives we identified that have provided or could provide duplicative support. As noted above, these initiatives provided almost all the federal financial support we identified related to wind in fiscal year 2011.

[57] One exception to this, according to Treasury officials, is Treasury's Qualifying Advanced Energy Project Credit, under which there is a limit on the total amount of support available. According to officials, DOE evaluates and ranks projects based on several criteria including their potential to create domestic jobs and reduce air pollution.

[58] As we have previously reported, federal agencies conduct similar assessments to guard against the potential for duplication or excessive support provided through subsidies for other types of projects. In particular, developers of affordable housing awarded multiple sources of public funding often must undergo subsidy layering reviews, which are a statutory requirement to assure that federal resources are neither duplicative nor wasteful when applied to affordable rental housing. See GAO, *Housing Assistance: Opportunities Exist to Increase Collaboration and Consider Consolidation*, GAO-12-554 (Washington, D.C.: Aug. 16, 2012).

[59] We have previously reported in another context that relying exclusively on applicants' self-certification of eligibility does not prevent ineligible firms from receiving program support. See GAO, *Service-Disabled Veteran-Owned Small Business Program: Vulnerability to Fraud and Abuse Remains*, GAO-12-697 (Washington, D.C.: Aug. 1, 2012).

[60] GAO, *Standards for Internal Control in the Federal Government,* GAO/AIMD-00-21.3.1 ("Green Book") (Washington, D.C.: November 1999).

[61] It was beyond the scope of our work to independently analyze the financing of specific wind projects in order to corroborate estimates of investors' returns or to determine whether or not federal support for such projects was necessary in order for the projects to be built. The estimates of returns and statements related to the projects' financial viability presented here do not necessarily indicate whether developers or investors would have committed their financial resources to these projects without the accompanying federal support. In commenting on a draft of this report, DOE noted that the Oregon wind project was of a significantly larger scale than other wind projects that had recently been financed, and that DOE's support mobilized a group of private investors to enter into a risk-sharing partnership with the federal government at a time when long-term capital for large-scale renewable energy projects was scarce.

[62] According to DOE officials, this estimate represents the internal rate of return for the private equity invested in the project and is based on several favorable assumptions. The officials

said these assumptions—including high quality wind resources and no equipment malfunctions or construction delays—could have changed since the estimate was developed in September 2010, reducing the actual rate of return for the project.

[63] Treasury does not have discretion in awarding these grants; taxpayers who are eligible for support under the Internal Revenue Code are entitled to that support.

[64] The American Taxpayer Relief Act of 2012, Pub. L. No. 112-240, 126 Stat. 2313 (2013), retroactively extended the deadline for wind facilities to claim the PTC by one year, from December 31, 2012 to December 31, 2013. In addition, it changed the definition of eligible facilities to those that begin construction prior to the deadline, rather than those that are placed in service prior to the deadline. Although the PTC was retroactively extended, according to financial professionals, a slowdown in wind deployment is expected in 2013 because wind project developers were uncertain about the future status of the PTC. As it can take 18 to 24 months to develop a new wind farm, investment decisions made in 2012 will impact future deployment levels.

[65] Lawrence Berkeley National Laboratory, *Preliminary Evaluation of the Impact of the Section 1603 Treasury Grant Program on Renewable Energy Deployment in 2009* (Berkeley, CA: April 2010).

[66] In addition, federal support for the construction of wind and other renewable energy projects is important because, according to studies by DOE's national laboratories, these projects are capital-intensive to build but have no ongoing fuel costs. In contrast, fossil fuel projects are less capital-intensive (per unit of production) but have higher fuel costs. Also, whereas federal tax incentives for fossil fuel projects are often distributed throughout the fuel cycle (e.g., from exploration and extraction to transportation and power production), tax incentives for wind projects are generally used to finance capital and generation costs.

End Notes for Appendix I

[1] GAO-12-260.

[2] For purposes of GAO-12-260 and, therefore, this report, we defined an initiative as a program or group of agency activities serving a similar purpose or function. For purposes of this report, we considered initiatives to be wind-related if they could or did promote or enable wind energy development, either exclusively or as part of a broader initiative. For instance, we considered initiatives with a focus on broader electricity issues—such as transmission or grid integration activities—to be wind-related if they encouraged or promoted wind energy, even if doing so was not an explicitly stated goal of the initiative. Similarly, for initiatives such as basic science research programs that did not directly promote development of wind-specific technologies, we considered them to be wind-related if they included research that could promote the future development of wind energy.

[3] For purposes of GAO-12-260, research and development includes efforts that further the knowledge of or the ability to create, develop, or improve technologies, including technology demonstrations; commercialization and deployment includes activities that provide incentives for the implementation or promote the competitiveness of renewable energy technologies in the commercial marketplace; and regulation, permitting, and compliance include activities that ensure compliance with federal laws and regulations regarding renewable energy production, use, technologies, or facilities. In addition, we included some initiatives categorized under the "other" federal role—which includes other

types of efforts related to renewable energy that, for example, provide information, services, or other types of support—if the initiatives included some specific focus on promoting the development of wind energy.

[4] For the purposes of this report, we considered an initiative "active" if it was authorized, planned, funded, or implemented in the fiscal year. For example, if funds for an initiative were obligated in a previous fiscal year, but its activities were still ongoing in fiscal year 2011, we considered it to be active in fiscal year 2011. Similarly, if funds for an initiative had not yet been obligated but the initiative was authorized or in the planning stages in fiscal year 2011, we considered it to be active in fiscal year 2011.

[5] Tax expenditures are reductions in federal tax liabilities that result from provisions of the federal tax laws that (1) allow a special exclusion, exemption, or deduction from gross income or (2) provide a special credit, preferential tax rate, or deferral of tax liability. Tax expenditures result in revenue losses for the federal government, which forgoes some of the tax revenues that it would have otherwise collected, while the taxpayers that take advantage of the tax expenditures pay lower taxes than they would otherwise have to pay. See also GAO-05-690.

[6] Most wind-related initiatives supported a range of energy sources in addition to wind. We asked agency officials to provide actual or estimated obligations data for their initiative's activities that were specifically related to wind.

[7] When multiple initiatives provide financial support to the same recipient for a single project, we consider that financial support to be duplicative. However, for the purposes of this report, we do not consider the initiatives themselves to be duplicative if they also support other recipients aside from those who have received duplicative financial support.

[8] In addition, we categorized certain recipients as "other" if they did not fit under any of the four main categories. Such recipients included federal agencies that develop or regulate wind resources, or private entities that did not fit under the energy industry or researcher categories.

End Notes for Appendix III

[1] The majority of wind-related initiatives, including all tax expenditures listed here, supported a range of energy sources in addition to wind.

In: Wind Energy
Editor: Anas Desmet
ISBN: 978-1-62808-069-8

Chapter 2

U.S. WIND TURBINE MANUFACTURING: FEDERAL SUPPORT FOR AN EMERGING INDUSTRY*

Michaela D. Platzer

SUMMARY

Increasing U.S. energy supply diversity has been the goal of many Presidents and Congresses. This commitment has been prompted by concerns about national security, the environment, and the U.S. balance of payments. Investments in new energy sources also have been seen as a way to expand domestic manufacturing. For all of these reasons, the federal government has a variety of policies to promote wind power.

Expanding the use of wind energy requires installation of wind turbines. These are complex machines composed of some 8,000 components, created from basic industrial materials such as steel, aluminum, concrete, and fiberglass. Major components in a wind turbine include the rotor blades, a nacelle and controls (the heart and brain of a wind turbine), a tower, and other parts such as large bearings, transformers, gearboxes, and generators. Turbine manufacturing involves an extensive supply chain. Until recently, Europe has been the hub for turbine production, supported by national renewable energy deployment policies in countries such as Denmark, Germany, and Spain. However,

* This is an edited, reformatted and augmented version of Congressional Research Service, Publication No. R42023., dated January 16, 2013.

support for renewable energy including wind power has begun to wane across Europe as governments there reduce or remove some subsidies. Competitive wind turbine manufacturing sectors are also located in India and Japan and are emerging in China and South Korea.

U.S. and foreign manufacturers have expanded their capacity in the United States to assemble and produce wind turbines and components. About 470 U.S. manufacturing facilities produced wind turbines and components in 2011, up from as few as 30 in 2004. An estimated 30,000 U.S. workers were employed in the manufacturing of wind turbines in 2011. Because turbine blades, towers, and certain other components are large and difficult to transport, manufacturing clusters have developed in certain states, notably Colorado, Iowa, and Texas, which offer proximity to the best locations for wind energy production. The U.S. wind turbine manufacturing industry also depends on imports, with the majority coming from European countries, where the technical ability to produce large wind turbines was developed. Although turbine manufacturers' supply chains are global, recent investments are estimated to have raised the share of parts manufactured in the United States to 67% in 2011, up from 35% in 2005-2006.

The outlook for wind turbine manufacturing in the United States is more uncertain now than in recent years. For the past two decades, a variety of federal laws and state policies have encouraged both wind energy production and the use of U.S.-made equipment to generate that energy. A continuing challenge for the industry is uncertainty about one main federal policy tool in the deployment of wind power, the production tax credit (PTC), which Congress has extended eight times and let lapse on four occasions. Most recently, the PTC expired at the end of 2012, but a few days later, Congress extended it through year-end 2013. At least a dozen wind turbine manufacturers announced layoffs or hiring freezes at U.S. facilities in 2012, citing concern about the PTC's future as one reason. Other factors affecting the health of the U.S. wind industry are intense price competition from natural gas, an oversupply in wind turbines, and softening demand for renewable electricity.

INTRODUCTION

This report discusses the U.S. wind turbine manufacturing industry, its supply chain, employment and international trade trends, major federal policy efforts aimed at supporting the industry, and issues affecting its future. The wind industry's national trade group, the American Wind Energy Association (AWEA), reported an estimated 30,000 Americans were employed directly and indirectly in wind turbine manufacturing in 2011, compared to 2,500 in

2004. Another 45,000 U.S. workers reportedly were employed in other parts of the wind industry in 2011, including construction and services.[1] Wind turbine equipment and component manufacturing jobs range in pay from about $30,000 to around $90,000, according to the Bureau of Labor Statistics.[2] Following an unprecedented period of growth in the U.S. wind power market between 2005 and 2009, about half as many new wind turbines were installed in 2011 (some 3,500) as in 2009.

Aside from GE Energy and Clipper Windpower, most of the manufacturers that sell, assemble, or manufacture turbines and wind-related components in the U.S. market are headquartered outside the United States. Vestas, Gamesa, and Siemens are among the European manufacturers that have responded to government regulations that mandate the use of renewables, including wind power. Other firms manufacturing wind turbines for the U.S. wind market include Japanese and Indian companies such as Mitsubishi and Suzlon. Manufacturers from South Korea and China are also expanding production capacity and entering the U.S. market.

Federal interest in the U.S. wind turbine manufacturing industry is based on (1) increasing the role of clean energy technology in energy production; (2) encouraging advanced manufacturing and the creation of skilled manufacturing jobs; and (3) enhancing the diversity of U.S. energy sources.[3]

Wind energy, like many energy technologies, benefits from government incentives.[4] Without them, it does not appear likely that there would be a U.S. wind turbine industry. To a large extent, the federal government sets the framework and influences the pace of domestic wind power development.

One of the main federal policy tools to encourage wind generation is a tax credit, known as the production tax credit (PTC), which is slated to expire at the end of 2013.[5] Other policy drivers include state renewable portfolio standards, which have been adopted by more than half the states to mandate production of electricity from "clean" sources.[6] No nationwide renewable electricity standard currently exists, but the Obama Administration and some Members of Congress have endorsed the concept.[7] These policies do not directly address manufacturing, but greater wind power adoption supports the development of a U.S. wind energy manufacturing base. In addition, the federal government and some state governments have maintained programs that provide financial incentives for manufacturing of wind power equipment.

Many international wind turbine manufacturers and component suppliers have opened manufacturing facilities in the United States since 2005. In 2011, there were more than 470 U.S.- based wind turbine manufacturing facilities—a 10-fold increase in five years—ranging from wind turbine assembly plants to factories producing various wind-related components including large bearings, castings, electrical wiring, fasteners, hydraulics, and power electronics. Only a small number of these factories are dedicated exclusively to building turbine parts (blades, towers, and nacelles); the others manufacture components for various industrial uses, including wind-specific products. Given the interest in wind power around the world, manufacturers with U.S. production facilities may be able to increase exports of advanced wind-energy components. Around $250 million in fully assembled wind turbines were exported from the United States in 2011.

The industry's future in the absence of government support, however, is open to question. While the cost of electricity from land-based wind turbines is less than the cost of power from other alternative sources, such as concentrated solar plants and geothermal installations, it is still, in general, somewhat higher than the cost of power from new gas-fired generators. This means that without government support, electricity suppliers' demand for wind turbines would be relatively limited. It is possible that, if existing policy tools are allowed to expire, wind industry manufacturing will face a difficult future. On the other hand, it is imaginable that technological improvements in wind generation and higher costs for construction of fossil-fuel power plants could at some point make wind cost-competitive with coal and gas as a source of electricity, creating a brighter outlook for wind turbine manufacturing.

Wind Turbine Manufacturing

Wind turbine manufacturing is at the core of the multifaceted wind power industry. Because of the use of castings, forgings, and machining, turbine manufacturing is a significant contributor to U.S. heavy manufacturing. By the end of 2011, more than 38,000 wind turbines were installed in the United States.[8] Procurement of wind turbines accounts for an estimated 60% to 70% of overall expenses for wind energy developers.[9]

The market potential of offshore wind power is not covered in this report. No offshore projects have been installed in the United States to date, and the industry faces difficulties with permitting, financing, and infrastructure availability.[10] So far, Cape Wind, off the coast of Nantucket in Massachusetts, is the only project that has a commercial wind energy development lease with the U.S. government.[11] AWEA reports that at the end of 2011, there were 15 different proposed offshore wind projects in the United States, and a proposed offshore transmission line.[12] To help spur offshore wind development, in

December 2012, the Department of Energy (DOE) awarded $28 billion in grants to seven offshore wind projects in six states, with the aim of three to begin commercial operations by 2017.[13] Also, this report does not cover small wind turbine manufacturing, which AWEA defines as turbines with rated capacities of 100 kilowatts (kW) or less. This segment of the wind turbine market appears to be growing. According to the World Wind Energy Association, worldwide more than 330 manufacturers offer small wind turbines.[14] AWEA's most recent data indicate that 95 manufacturers of small wind turbines were based in the United States in 2009.[15]

Historical Overview

The use of a wind turbine to generate electricity is an American invention of the late 19th century.[16] The development of U.S. commercial wind turbine manufacturing can be traced back to the 1970s, when the U.S. government advanced the technology in response to the oil crises of 1973 and 1979 as an alternative to power generation from fossil fuels.

The first U.S. wind farms were developed in California, an early adopter of policies favorable to wind energy, and the state dominated worldwide wind development in the early 1980s.[17] This created a market for wind turbine manufacturers. Enertech, U.S. Windpower (renamed Kenetech in 1988), and Zond were among the American suppliers. Other U.S. manufacturers included technology and aerospace firms such as Westinghouse and Boeing. In 1986, 60 U.S. firms produced turbines for the California market.[18] Foreign suppliers from Denmark, Germany, Japan, and the Netherlands, among other countries, also sold their wind turbines in California.[19] The California "wind rush" became the training ground for several firms, including the Danish manufacturer Vestas, now the world's largest manufacturer of utility-scale wind turbines.[20]

However, a drop in oil prices, along with reductions in government tax credits, caused a near total collapse of this market in the mid-1980s.[21] By the end of the decade, many wind turbine manufacturers went bankrupt as the industry adjusted to a much smaller market.

For the next two decades fuel prices were low and U.S. incentives spotty. In the United States, annual installed wind power capacity slowed from 1987 to 2000. The entire U.S. wind fleet exceeded 1,000 megawatts (MW) for the first time in 1986, but then took 13 years to reach approximately 2,400 MW.[22]

In the 1990s a more sustained market for wind power and wind turbine manufacturing evolved overseas. Strong, consistent government incentives and policies, which have included a policy mix of direct government investment, tax breaks, loans, regulations and laws that cap or tax emissions, supported the development of manufacturers abroad, particularly in Europe.[23] This allowed wind turbine manufacturers to establish themselves in countries such as Denmark, Spain, and Germany, where many wind turbine manufacturers are now based.[24]

Demand for Wind Turbines and Components

Demand for wind turbines and components is driven by growth in wind power capacity. More consistent U.S. policies have resulted in a substantial increase in cumulative utility-scale wind power capacity, from 9,000 MW in 2005 to more than 46,000 MW in 2011.[25] The United States was second to China in cumulative and new installed wind power capacity in 2011.[26] China and the United States accounted for more than 45% of total installed worldwide wind power capacity at the end of 2011.[27] The size of the U.S. market has made the United States an attractive investment location for wind turbine and wind component manufacturers. However, the prospects for 2013 and beyond are clouded owing to several factors, including the fate of the PTC, low-cost natural gas, and manufacturing overcapacity in the wind turbine sector.[28]

Major customers for wind turbine manufacturers are large independent power producers (IPPs) and utilities such as Iberdrola Renewables, NextEra Energy Resources, Horizon-EDPR, Terra-Gen, Duke Energy, or Xcel Energy, which purchase wind turbines for commercial electricity generation.[29] Other wind turbine customers include universities and military bases, but these customers account for a very small share of the market.

Commercial utility-scale onshore wind turbines are installed at wind farms, which are clusters of wind turbines grouped together to produce large amounts of electricity. Currently, there are some 975 wind farms in the United States.[30] The largest is located in California, and there are several huge wind farms in Texas (see Table 1), which is by far the leading state in wind energy output, with over 10,000 MW of installed capacity at year-end 2011. Other large wind-power projects are in Indiana, Oregon, and Iowa. Several large U.S. wind farms are owned and managed by overseas companies. For example, the Roscoe, TX, wind farm is owned and operated by Germany-based E.ON Climate and Renewables. It consists of more than 600 wind turbines purchased from three different manufacturers: Mitsubishi, General Electric (GE), and Siemens.

Table 1. Largest U.S. Wind Power Projects

Project Name	State	Installed Capacity (MW)	Year Online	Owner	Number of Turbines/ Manufacturer(s)
Alta Wind Project	California	981.0	2010, 2011	Terra-Gen Power	377/Vestas
Roscoe	Texas	781.5	2008	E.On Climate & Renewables	627/Mitsubishi, Siemens, GE
Horse Hollow	Texas	735.5	2006, 2006	NextEra Energy Resources	421,GE/Siemens
Capricorn Ridge	Texas	662.5	2007, 2008	NextEra Energy Resources	407,Mitsubishi/ Siemens
Sweetwater	Texas	585.3	2003, 2005, 2007	Babcock & Brown Wind, Catamount	392/Vestas,GE/ Siemens

Source: American Wind Energy Association (AWEA), U.S. Wind Industry Annual Market Report, 2011, p. 32.

Wind Turbine Suppliers

International Manufacturers Dominate Wind Turbine Manufacturing

In 2011, 10 wind turbine manufacturers accounted for 85% of the global market measured by newly installed capacity. The three largest manufacturers were:

- Vestas at 12.9% (Denmark);
- Goldwind at 9.4% (China); and
- GE at 8.8% (U.S.).[31]

Other leading manufacturers are listed in Appendix A. These firms are headquartered in Europe, the United States, India, and China. GE Energy[32] and UTC/Clipper Windpower[33] are the only U.S.-headquartered utility-scale wind turbine manufacturers.

Some manufacturers, including Gamesa, Vestas, and Suzlon, focus exclusively on wind turbines. Others are part of larger diversified companies. All pursue a global business strategy, which means selling outside their home markets. Many operate manufacturing facilities throughout the world, including the United States, Europe, and China.

Recently, several Chinese companies have begun producing wind turbines, selling mainly in the large and growing China market.[34] China, which had virtually no wind turbine manufacturing capabilities in 2005, is now home to over 270 producers,[35] some of them capable of producing complete wind turbine systems with locally made products.[36] Four of the top 10 manufacturers worldwide in 2011 were headquartered in China (see Appendix A), where, by some estimates, turbines can be manufactured for 30% less than in Europe, the United States, or Japan.[37] Some Chinese firms apparently are looking for overseas markets,[38] but concerns about the quality of Chinese turbines are one factor that might limit foreign sales since Chinese-made turbines are not yet seen as being as high in quality as European and American ones.[39]

South Korean companies are also making huge investments in wind turbine production. Two large South Korean shipbuilders, Hyundai Heavy Industries and Samsung Heavy Industries, have announced their intention to manufacture wind turbines. Other Korean firms undertaking wind turbine technology development include Daewoo, Hanjin, STX, Rotem, and Unisom. Additionally, South Korean wind turbine component manufacturers like Doosan, Hanjin, Taewoong, Hyosung, CS Wind, and Korea Tech are becoming important suppliers of towers, blades, generators, transformers, gearboxes, nacelle control systems, and cables.[40]

U.S. Market Attracts More Foreign Wind Turbine Manufacturers

The leading manufacturers of utility-scale wind turbines in the United States are shown in Table 2. In 2011, nearly two dozen wind turbine manufacturers—a five-fold increase in six years— installed nearly 3,500 new turbines nationwide, generating 6,800 MW of new capacity.[41] This was down from the 2009 peak, when some 5,700 new wind turbines were installed, adding nearly 10,000 MW of new utility-scale wind capacity.[42] Since the inception of utility-scale wind energy production, U.S. electric generators have installed more than 40,000 turbines with approximately 52,000 MW of capacity. In 2011, GE continued to lead in the number of new wind turbine installations, although its market share has declined over time.[43]

Table 2. Annual Wind Turbine Installations in the United States Top 10 Manufacturers by Selected Years, Ranked by Number Installed in 2011

Original Equipment Manufacturer (OEM)/Assembler[a]	Location of Headquarters	2005 (# of Turbines)	2009 (# of Turbines)	2010 (# of Turbines)	2011 (# of Turbines)
GE Energy	United States	954	2,663	1,679	1,252
Vestas	Denmark	403	830	75	952
Siemens	Germany	0	505	360	534
Suzlon[b]	India	8	344	201	159
Mitsubishi	Japan	190	491	146	133
Nordex	Germany	0	25	8	115
Clipper	United States	1	242	28	103
REpower	Germany	0	165	34	84
Gamesa	Spain	25	300	282	77
Alstom	France	0	0	0	25
All Others		33	200	129	30
Total		1,614	5,765	2,942	3,464

Source: AWEA, U.S. Wind Industry Annual Market Report, 2009, 2010, and 2011. The number of turbines is based on data compiled by AWEA and is accurate as of December 17, 2012, but is subject to revision.

[a] An OEM designs the turbine, typically assembles the nacelle, and sells the completed unit to developers.

[b] Suzlon acquired 100% of REpower during 2011. The two firms jointly had 243 installations in 2011.

WIND TURBINE COMPONENTS, RAW MATERIALS, GLOBAL SUPPLY CHAIN, AND U.S. MANUFACTURING CAPACITY

Wind Turbine Components

A wind turbine is a collection of operating systems that convert energy from wind to produce electricity. Utility-scale wind turbines are massive, complex pieces of machinery which come in many sizes and configurations. Wind turbine blades range in size from 34 to 55 meters, the hub can weigh 8 to 10 tons, and towers are usually 80-100 meters tall and weigh 55 to 70 tons. According to AWEA, the installation of over 5,700 turbines in the United

States in 2009 required industrial manufacturers to supply 17,000 blades and tower sections, approximately 3.2 million bolts, 36,000 miles of rebar, and 1.7 million cubic yards of concrete.[44] In simple terms, as shown in Figure 1, the major components in a wind turbine consist of:

- a rotor comprising four principal components—the blade, the blade extender, the hub, and the pitch drive system;
- a nacelle, the external shell or structure resting atop the tower containing and housing the controller, gearbox, generator, large bearings, connecting shafts, and electronic components that allow the turbine to monitor changes in wind speed and direction;
- a tower, normally made of rolled steel tube sections that are bolted together to provide the support system for the blades and nacelle; and,
- other components, including transformers, circuit breakers, fiber optic cables, and ground-mounted electrical equipment.[45]

Beyond the major components, there are many subcomponents in a wind turbine. The percentages shown in Figure 2 indicate the costs of the components relative to the overall cost of a turbine. The tower, for example, is over 26% of the total cost of a wind turbine, rotor blades 22%, the gearbox 13%, and the other components 5% or less.

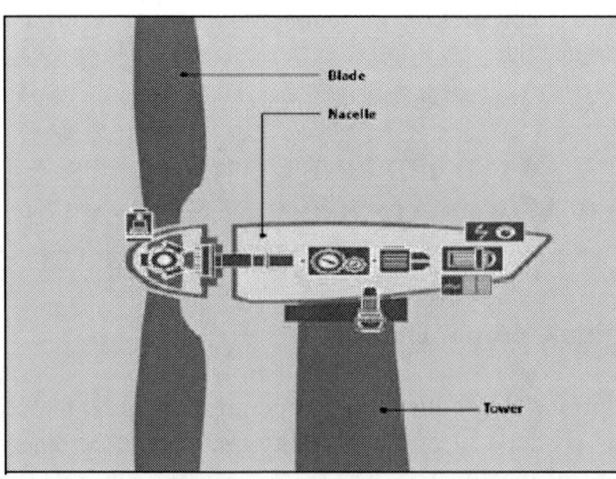

Source: Wind Directions, *"Supply Chain: The Race to Meet Demand,"* January/February 2007.

Figure 1.Wind Turbine Overview.

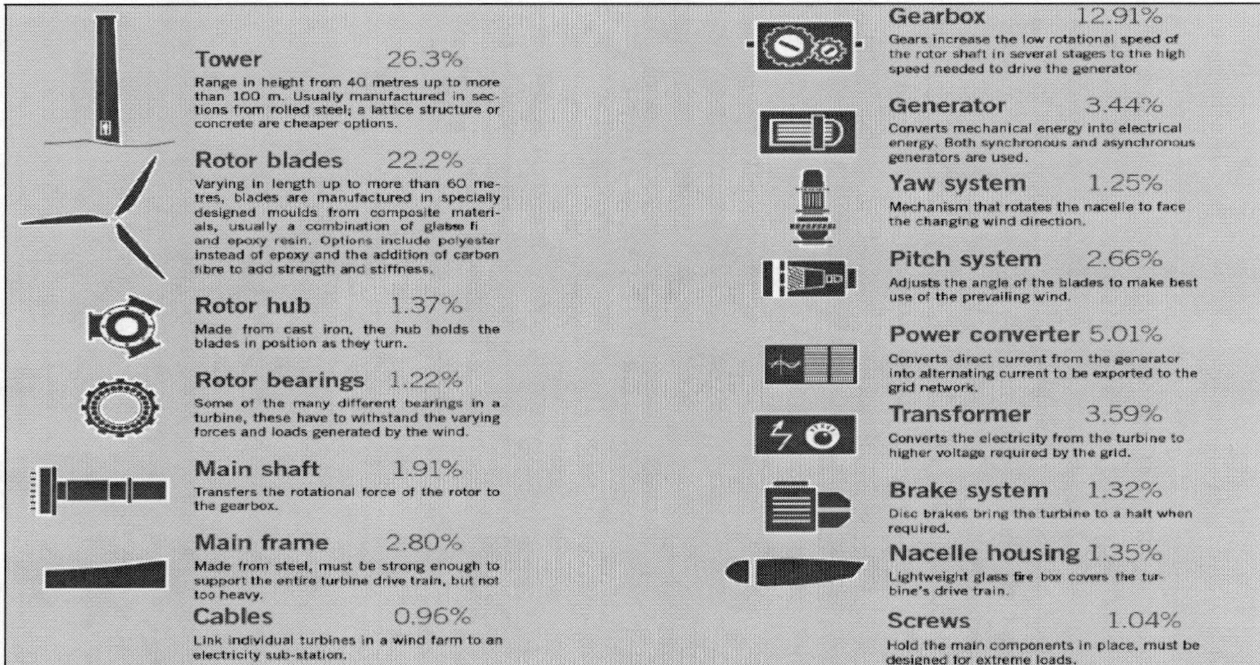

Source: Wind Directions, *"Supply Chain: The Race to Meet Demand,"* January/February 2007.

Figure 2. Wind Turbine Components; Contribution of main parts as a percentage of overall costs based on a REpower MM92 Turbine.

Wind turbines vary greatly in size and are getting larger as technology advances. They have grown from dozens of kilowatts in the early 1980s to as large as 7 MW.[46] Most land-based wind turbines are in the 1.5 MW to 3 MW range.[47] Components also change as technology improves. European and U.S. wind turbine manufacturers have invested heavily over the decades in developing their respective turbine technologies, leading to improvements in the efficiency of wind blades and turbines and longer turbine life. New wind turbine manufacturers, especially from China, are not yet globally competitive. According to recent research, they generally lack state-of-the-art technology, focus mainly on producing smaller turbines, and experience significant quality control problems.[48] Raw materials availability and changing commodity prices of raw materials used in wind turbines affect production costs.[49] A typical wind turbine is made primarily of steel (about 90% by weight) (see Table 3). Aluminum and other light-weight composites are also important, particularly for blade manufacturing. Other core materials include pre-stressed concrete, copper, and fiberglass. Turbines also utilize permanent magnets, cast iron, carbon fiber, rubber, epoxy, ferrite, brass, ceramics, and Teflon.[50]

Table 3. Raw Materials Requirements for Wind Turbines based on a 1.5MW Wind Turbine by % of Weight, including blades and towers

	Steel	Fiberglass	Copper	Concrete	Adhesive	Aluminum	Core Materials
Weight %	89.1%	5.8%	1.6%	1.3%	1.1%	0.8%	0.4%

Source: U.S. Department of Energy, 20% Wind Energy by 2030, p. 63, July 2008.

Global Wind Turbine Assembly Supply Chain

Wind turbines are manufactured by original equipment manufacturers, or OEMs, which design, assemble, and brand their products. Similar to automobile assemblers that make a car or truck, OEMs are mostly system integrators. Assemblers must bring together an estimated 8,000 precision parts and components to produce a wind turbine.[51] One supplier might roll large plates of steel into the towers that support the turbine. A second company might make the turbine blades from special carbon fiber materials, and a third might manufacture the electronic computerized control systems. Each of these components might be produced domestically, might be assembled domestically from imported inputs, or might be imported as an assembled product.[52]

Tier 1 and Tier 2 Wind Turbine Component Suppliers

Many suppliers and specialty firms are part of this complex global supply chain. Tier 1 suppliers make large components such as towers, hubs, blades, or gearboxes. They include firms such as LM Wind (blades), SKF (bearings), and Winergy (gearboxes). Tier 2 suppliers produce subassemblies such as ladders, fiberglass, control systems, hydraulics, power electronics, fasteners, resin, machine parts, or motors. They include companies such as American Roller Bearings (power transmission bearings), Cardinal Fasteners (structural fasteners), and Timken (power transmission bearings).

Manufacturing Strategies

A wind turbine is a significant investment. Researchers at the Lawrence Berkeley National Laboratory reported that wind turbine transaction price quotes can range from as low as $900/kilowatt (kW) to a high of $1,400/kW,[53] meaning that an average 2 MW turbine would cost between $1.8 million and $2.8 million, plus installation costs.

Each wind turbine assembler uses different sourcing strategies and levels of vertical integration. Some produce almost all major components internally or through subsidiaries, while others outsource many of their critical components.[54] For instance, some manufacturers produce blades, generators, or gearboxes in-house, while others opt for outside suppliers. Hundreds of smaller companies make specialized parts such as clutches, rotor bearings, fasteners, sensors, and gears for the wind industry.[55] Illustrative examples of some of the thousands of components in a modern wind turbine are shown in Table 4.

Very high levels of expertise and specialization are required of wind turbine suppliers, with the level of precision similar to that of the aerospace industry. Turbine manufacturers often establish relationships with suppliers in the interest of quality, as a failure in a turbine part can be very expensive to fix. Wind turbines are expected to survive largely unattended in extreme climactic conditions for a design life of as much as 20 years.[56] Product quality is also of concern to wind farm operators, as a malfunctioning turbine can reduce operating revenue.[57]

Table 4. Selected Wind Turbine Components

Towers:	**Nacelle:**	**Foundation:**
• Towers	• Nacelle Cover	• Rebar
• Ladders	• Nacelle Base	• Concrete
• Lifts	• Heat exchanger	• Casings
Rotor:	• Controllers	Other:
• Hub	• Generator	• Transformers
• Nose Cone	• Power Electronics	• Bolts/Fasteners
• Blades	• Lubricants	• Wire
• Pitch Mechanisms	• Filtration	• Paints and Coatings
• Drives	• Insulation	• Lighting Protection
• Bakes	• Gearbox	• Steelworking/Machining
• Rotary Union	• Pump	• Communication Devices
	• Drivetrain	• Control and Condition Monitoring Equipment
	• Ceramics	• Electrical Interface and Connections
	• Shaft	• Batteries
		• Bearings
		• Brakes

Source: AWEA, *Manufacturing Supplier Handbook for the Wind Energy Industry*, 2011, p. 29. http://www.awea.org/ issues/supply_chain/upload/Supplier-Hand book.pdf.

U.S. Wind Turbine Manufacturing Facilities

At the end of 2011, the American Wind Energy Association reported that more than 470 wind turbine manufacturing facilities were located in the United States, up substantially from the 30-40 wind-related manufacturing facilities nationwide in 2004. Over that period, the number of tower plants increased from 6 to 18; blade facilities rose from 4 to 12; and, nacelle assembly facilities grew from 3 to 14.[58] Total investment in facilities to manufacture for the wind industry in the United States has exceeded $1.5 billion.[59]

Greater demand for wind turbines, cost savings related to transportation, and concern about the risks associated with currency fluctuations are among the reasons wind turbine and component manufacturers have opened new production facilities in the United States since 2005.[60] Even with increased domestic production capacity, wind turbine assemblers source parts and components on a worldwide basis, reflecting the industry's global supply chain. Many wind manufacturers with production facilities in the United States also produce elsewhere, typically in Europe and Asia.

U.S.Wind-Related Manufacturing Facilities, Number of Facilities by Selected Categories, December 2011

Towers	18
Blades	12
Nacelle Assembly	14
Fasteners	21
Bearings	21
Castings	14
Gearboxes	7
Generators	1

Source: AWEA.

Towers and Blades

Towers and blades were among the first wind products manufactured in the United States because they are large, expensive, and difficult to transport.[61] Thus, manufacturers find it easier and less costly to fabricate near their installation point. Many tower manufacturers in the United States are American companies and include firms such as Ameron, Trinity Structural Towers, DMI Industries, and Broadwind Towers. Foreign manufacturers, such as Gamesa and Vestas, also have located tower manufacturing facilities in the United States. In 2012, several major tower producers, including Otter Tail Corporation, Katana Summit, and DMI Industries, exited the sector or announced plans to scale back production. Suppliers of blades have increased their U.S. manufacturing capacity, with three times as many facilities in 2011 as in 2005. For example, LM Wind Power, headquartered in Denmark, is the largest supplier of blades in the world; it now produces blades at two U.S. manufacturing facilities.[62] Other blade manufacturers with U.S. production facilities include two American companies, TPI Composites and Molded Fiberglass. Both make blades for GE.

Turbine Nacelle Assembly

European OEMs, including Gamesa, Nordex, Siemens, and Vestas, have opened nacelle assembly plants in the United States in recent years.[63] Some started investing in the United States heavily after the American Recovery and Reinvestment Act (P.L. 111-5) passed in 2009.[64] Siemens and Nordex also opened their first U.S. nacelle assembly facilities in 2009. GE has three nacelle assembly facilities in the United States, all established prior to 2005, and also operates turbine component plants in China, Vietnam, and Europe. With the exception of DeWind, which was a German-owned manufacturer acquired by South Korea's Daewoo Shipbuilding & Marine Engineering Company in 2009, Asian manufacturers lag in establishing a U.S. nacelle manufacturing presence. Japanese-headquartered Mitsubishi had expected to open its first U.S. nacelle assembly plant in 2012, but has delayed the opening of that facility.[65] Other manufacturers, such as Suzlon, which idled its turbine rotor blade plant in Minnesota, have reduced their U.S. manufacturing presence.[66] Appendix B provides an overview of the varied investment strategies pursued by foreign-based wind turbine assemblers in the United States.

Other Wind Turbine Components

A more robust domestic manufacturing base for wind turbine components such as bearings, gearboxes, and power transmissions is also being established in the United States, albeit more slowly than for towers, blades, and nacelle assembly. Gearboxes and bearings are among the most critical components for any wind turbine manufacturer because failures in either of these parts mean the wind turbine will fail. Bearings for wind turbines are made by a few manufacturers, such as German-headquartered FAG[67] and U.S.-headquartered Timken.[68] Both have production capacity in the United States and operate factories in Europe and Asia. Gearboxes are also made by a relatively small number of companies, such as Winergy (now part of Siemens), which established U.S. production capacity in Illinois in 2009.[69] Winergy also makes gearboxes in Europe, China, and India.[70] Manufacturers of power transmissions, power converters, composite coatings, and sensors have also located wind-related production facilities in the United States.

Outlook

Falling natural gas prices, and continued uncertainty over the long-term future of the PTC, may have diminished manufacturers' interest in establishing more wind-related production facilities in the United States. An analysis by the National Renewable Energy Laboratory (NREL) reported that in 2011, 16 new turbine and component manufacturing facilities opened across the nation, compared to 13 in 2010.[71] In 2012, some manufacturers delayed implementing announced plans for new factories or expansion of existing ones, and several companies reduced their U.S. workforce.

An Emerging U.S. Wind Manufacturing Corridor

A concentration of tower, blade, and nacelle assembly plants is found in the central part of the United States, as shown in Figure 3. Texas, Iowa, Colorado, Arkansas, and Kansas are positioned near sites that are favorable for wind power generation, enabling manufacturers there to minimize transportation challenges and costs.

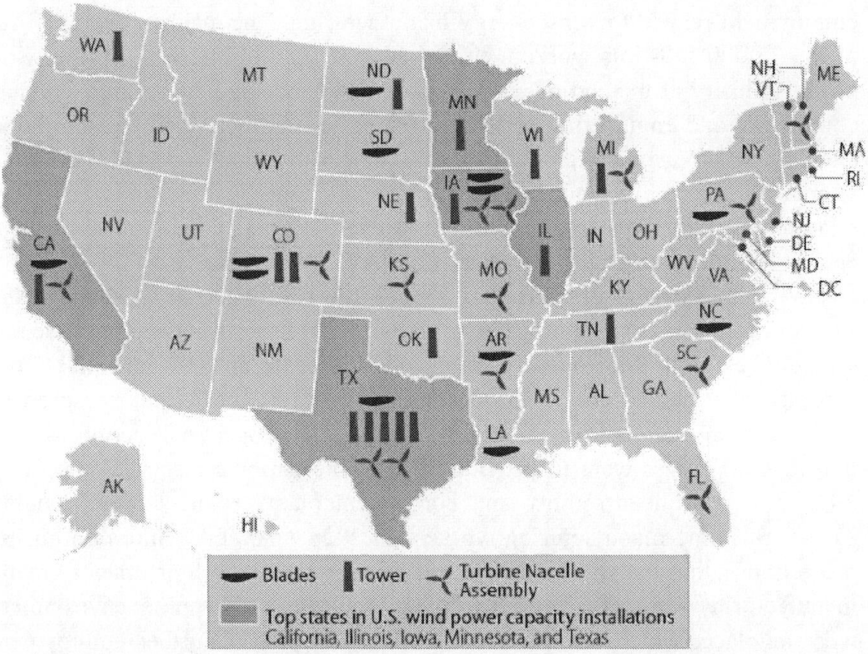

Source: CRS based on data from AWEA. The map shows 44 online tower, blade, and turbine nacelle assembly facilities at year-end 2011. It does not show the more than 400 facilities that produce wind components such as power transmissions, generators, gearboxes, or bearings.

Figure 3. Wind Turbine Manufacturing Facilities in the United States; By Tower, Blade, and Turbine Nacelle Assembly, 2011.

In addition, wind turbine assemblers and tower and blade manufacturers have been attracted to these states by incentive packages including property tax abatements, sales tax reductions, low-interest loans, and support for worker training. Other wind-related manufacturing facilities are located in Pennsylvania, Michigan, and Ohio, where the decline of automotive and heavy industrial manufacturing has left behind a workforce with prior experience with steel, assembly lines, robotics, and other aspects of heavy manufacturing.[72]

U.S. Wind Turbine Manufacturing Employment

In 2011, the wind turbine manufacturing sector supported an estimated 30,000 manufacturing jobs nationwide. This was only about one-fourth of U.S. employment related to wind energy manufacturing. The majority (some 60%) of the 75,000 full-time workers employed directly and indirectly in the wind power industry at the end of 2011 worked in finance and consulting services, contracting and engineering services, project development, and transportation and logistics.[73] About 4,200 jobs were in construction and 4,000 were in operations and maintenance. The number of manufacturing jobs has been relatively flat over the past three years, even as total employment in wind energy declined, according to figures from AWEA (see Figure 4).[74]

Wind turbine manufacturing is responsible for a very small share of the 11.7 million domestic manufacturing jobs in 2011, well under 1%. It seems unlikely, even if there were a substantial increase in U.S. manufacturing capacity, that wind turbine manufacturing will become a major source of manufacturing employment. In 2008, the U.S. Department of Energy forecast that if wind power were to provide 20% of the nation's electrical supply in 2030, U.S. turbine assembly and component plants could support roughly 32,000 full-time manufacturing workers in 2026.[75] AWEA's more optimistic projection is that the wind industry could support three to four times as many manufacturing workers as it does now if a long-term stable policy environment were in place, which implies a total of 80,000 jobs.[76] Further employment growth in the sector is likely to depend not only upon future demand for wind energy, but also on corporate decisions about where to produce towers, blades, nacelles, and their most sophisticated components, such as gearboxes, bearings, and generators.

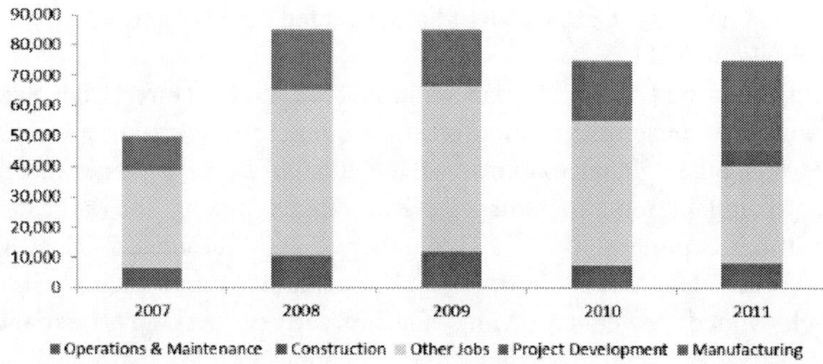

Source: AWEA, U.S. Wind Industry Annual Market Report, 2011.

Note: Other jobs include financial and consultant services, developers and development services, contracting and engineering services, and transportation and logistics. Project development employment is only available beginning in 2011.

Figure 4. Wind Energy Employment Trends; 2007-2011.

WIND TURBINE EQUIPMENT TRADE

U.S. Imports

As part of their global business strategies, wind turbine manufacturers continue to source a significant share of components outside the United States.[77] Imports of wind-powered generating sets, the main wind category covering fully assembled wind turbines and including other components such as blades and hubs when they are imported with the nacelle, grew from $482.5 million in 2005 to a peak of $2.5 billion in 2008. In 2009, imports of wind-powered generating sets dropped to $2.3 billion, then fell by another 46% to $1.2 billion, before rising by 1% in 2011 (see Figure 5).[78] An analysis of U.S. wind equipment trade by the U.S. International Trade Commission identified several explanations for the recent decline in U.S. imports of wind-powered generating sets, which include fewer wind turbine installations; decreasing prices; and the opening of new production facilities in the United States.[79]

The overwhelming majority (95%) of imported wind-powered generating sets come from Europe. In 2011, Denmark was the leading source of wind-powered generating sets, making up more than half (55%) of all imports into the United States. Italy, Germany, and Spain combined accounted for another

40% (see Figure 5). China and India accounted for 2% and 1% of imports, respectively, in 2011.

It appears that South Korean wind turbine manufacturers like Samsung, Hyosung, and Unison have ambitions to become leading exporters to the U.S. market and other global markets.[80] Even though China is home to 60 wind energy manufacturers, including several ranked among the largest in the world, it has exported only a small number of wind turbines, $351 million by value worldwide in 2011. However, Chinese manufacturers such as Goldwind, Sinovel, United Power, and Mingyang are actively seeking to expand their foreign sales. Between 2008 and 2011, 11 Chinese OEMs exported 194 wind turbines, based on one estimate, with the United States accounting for 59% of the installations.[81] Also, European turbine assemblers such as Vestas are now looking to open plants in China to supply the Chinese market, and possibly global markets.[82] Concerns about the quality of Chinese-made turbines and parts have prevented more rapid adoption of Chinese components. This may change as Chinese wind turbine products improve and as more foreign manufacturers establish operations in China.

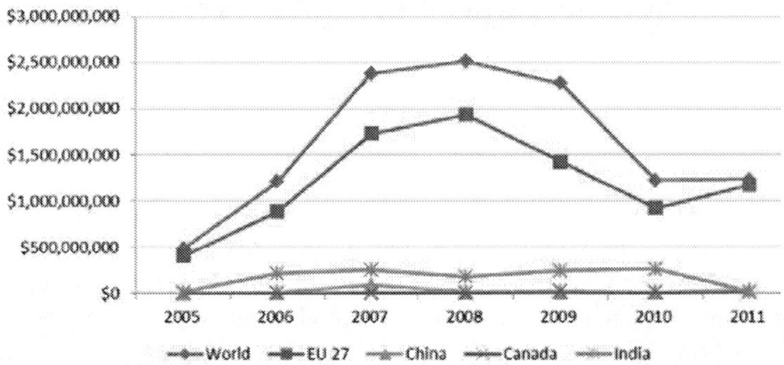

Source: Global Trade Atlas. These statistics only cover wind-powered generating sets (HS 8502.31), not components such as blades, towers or gearboxes imported separately.

Note: The import statistics are shown on a domestic consumption basis.

Figure 5. U.S. Imports of Wind-Powered Generating Sets, Select Countries; 2005-2011.

China's efforts to foster wind turbine manufacturing have been an irritant in the bilateral relationship. The United Steelworkers (USW) filed a claim in September 2010 that China's green technology policies are direct violations of China's World Trade Organization (WTO) obligations.[83] In June 2011, after the World Trade Organization panel upheld a U.S. complaint, the Office of the U.S. Trade Representative (USTR) announced that China will end a program of wind power equipment grants that required Chinese wind turbine manufacturers that received them to use domestic parts and components instead of foreign-made parts and components.[84]

Besides the USW complaint, the U.S. wind tower industry has been involved in an ongoing trade case. In December 2011, the Wind Tower Trade Coalition, representing four U.S. manufacturers of steel towers for wind turbines,[85] filed anti-dumping and countervailing duty (CVD) petitions with the U.S. Department of Commerce (DOC) and the International Trade Commission (ITC), alleging that Chinese and Vietnamese makers of wind towers have injured U.S. producers by selling their products in the United States at below-market prices. In May 2012, DOC ruled that Chinese exporters of utility scale wind towers are being unfairly subsided and announced preliminary CVD rates ranging from 13.74% to 26%.[86] In July 2012, DOC issued an affirmative preliminary anti-dumping ruling that could impose additional duties as high as 73% on Chinese towers imported into the United States.[87] Final determinations are scheduled for early 2013. If the dumping and subsidy cases lead to significant tariffs, the rulings may impact the magnitude and source countries of tower imports from China to the United States in future years. U.S. imports of other wind-related equipment, such as towers and blades, followed a similar pattern to wind-powered generating sets, with increases from 2005 to 2008 followed by a drop in 2009, then again in 2010, with a rise in 2011. But although more of these large components are being produced domestically, imports remain significant. China, Mexico, Vietnam, and South Korea were the main sources of imported towers and lattice masts in 2011.[88] China, Mexico, and Canada led in blade imports in 2011.[89] Some turbine components, such as bearings and gearboxes, are relatively easier to transport, and wind turbine assemblers might be more likely to continue to use global sourcing strategies for these less bulky components.

Domestic Content

Estimates indicate that U.S. content in recent years has increased to nearly 70% of the value of the average wind turbine installed in the United States.[90] In an August 2012 report, analysts at the Lawrence Berkeley National Laboratory calculated that the share of parts manufactured domestically nearly doubled from around 35% in 2005-2006 to 67% in 2011.[91] Public statements by major wind turbine assemblers appear to support the view that U.S.-made turbines now contain a larger share of domestic content than in previous years. For example, Gamesa reports that its domestic content on U.S.-made wind turbines is upwards of 65% and it has a local supply goal of 75%.[92] Vestas has stated domestic content in one class of its wind turbines has grown to 80%, and it expects to increase the overall percentage to 90%, including components and suppliers.[93] The *2011 Wind Technologies Market Report* notes "a growing amount of the equipment used in wind power projects has been sourced domestically in recent years. Whether that trend continues in the future may depend on the size and stability of the U.S. wind power market as well as the manufacturing strategies of emerging wind turbine manufacturers from Asia and elsewhere."[94]

U.S. Exports

Future growth of the U.S. wind turbine industry also depends on foreign markets. In 2010, the Obama Administration announced a goal to demonstrably increase renewable energy and energy efficiency exports like wind turbines.[95] Exports of wind-powered generating sets from the United States to the world remain relatively small, especially in comparison to imports, at only $255 million in 2011, up from $3.6 million in 2005 (see Figure 6).[96] U.S. producers may turn to foreign markets to offset falling domestic demand because of increasing market uncertainty and overcapacity in U.S. wind turbine equipment manufacturing.[97] The Western Hemisphere may be especially attractive to U.S.-based exporters of wind turbine equipment. For instance, the expansion of the Canadian and Mexican wind turbine markets could increase export opportunities for companies with manufacturing operations in the United States, including GE, Siemens, Gamesa, and Vestas.[98] Brazil is the largest market in Latin America for wind power, which could provide U.S.-producers of nacelles and wind subcomponents with fresh export opportunities. A counter-trend is that wind turbine assemblers also are localizing production in the large Brazilian market, including manufacturers like GE and Gamesa.[99] Although considerably smaller, there are other growing markets in Central and South America that could buy more U.S. wind products, including Honduras, Uruguay, and Chile.

If U.S. manufacturers begin to export more wind turbine equipment, they will have to contend with import tariffs, non-tariff barriers, and domestic industry subsidies. Tariff rates in some major markets are disproportionately higher than U.S. tariffs. For instance, the U.S. duty rate for wind-powered generating sets is 2.5%, compared to 14% in Brazil, 8% in China, 7.5% in India, and 2.7% in the European Union.[100] Subsidies and non-tariff barriers in major overseas markets like China are another potential constraint on U.S. exports.[101]

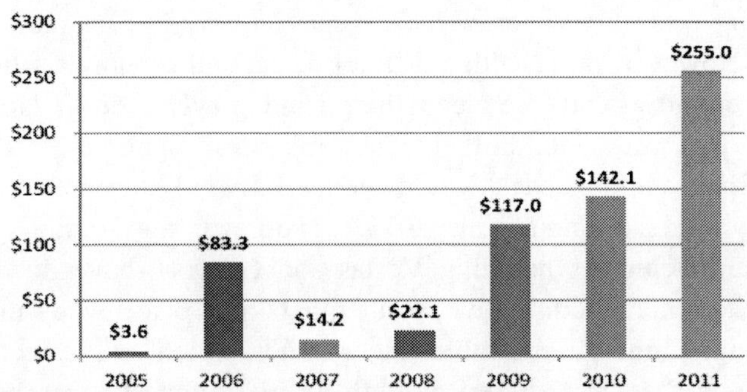

Source: Global Trade Atlas (accessed December 12, 2012).

Notes: These statistics only cover wind-powered generating sets, which refer to the complete nacelle and any items imported with the nacelle (HS 8502.31).

Figure 6. U.S. Exports of Wind-Powered Generating Sets; in millions of U.S. dollars, 2005-2011.

Several U.S. government programs are designed to encourage the export of renewable energy products, such as direct loans provided to wind manufacturers by the Export-Import Bank of the United States.[102] Owing to the Ex-Im Banks's environmental export financing program, for example, Clipper Windpower exported 27 wind turbines to Mexico in 2010 based on a direct loan from the Ex-Im Bank of $80.7 million.[103] In 2011 and 2012, Ex-Im also extended loans of $22 million for 55 Northern Power wind turbines to Italy,[104] $159 million for 51 Gamesa wind turbines to Honduras,[105] and $32 million for 55 LM Wind Power wind blades to Brazil.[106]

FEDERAL SUPPORT FOR THE U.S. WIND POWER INDUSTRY

Worldwide the wind power industry is driven by various types of government support, which range from tax credits to incentive policies like feed-in tariffs.[107] These incentives have been much larger in several foreign countries than in the United States, which has helped to spur the manufacturing of wind turbines in Europe and Asia. More recently, however, many countries— especially in Europe—have begun to reduce subsidies for renewables, including wind.[108] In Europe, feed-in tariffs[109] are among the policy tools that have been used to promote wind power, and have been credited by industry advocates like the European Wind Energy Association[110] with driving renewable energy growth, particularly in Denmark, Spain, and Germany. However, faced with a difficult fiscal and economic situation, some European countries have reduced their wind power feed-in tariffs and are taking a more critical look at their renewable energy policies.[111] For instance, in 2010, Spain announced it would reduce its wind subsidies by 35% from January 1, 2011, to January 1, 2013.[112] Some of the leading global wind turbine manufacturers, including Vestas and Gamesa, have downsized their operations to remain competitive, while others may place even more emphasis on exporting. China's Renewable Energy Law, which took effect in 2006, is one measure that has driven growth in the domestic market.[113] China introduced a feed-in tariff for wind power generation in 2009.[114] The Chinese government also implemented various policies to encourage the development of local manufacturing and technology development.[115] In the United States, various federal policies also have been instrumental in the development of a domestically based wind power sector, including

- the production tax credit (PTC)/Investment Tax Credit (ITC), which will expire at the end of 2013;
- an advanced energy manufacturing tax credit (MTC), which reached its funding cap in 2010 (no additional funds were allocated to continue with the MTC);
- the Section 1603 Treasury Cash Grant Program, which required that wind projects begin construction by December 31, 2011, and be placed in service by December 31, 2012; and
- the Section 1705 Loan Guarantee Program for commercial projects, which includes manufacturing facilities that employ "new or significantly improved" technologies.

The wind industry asserts that a national renewable electricity standard is needed to create longterm stability and to attract investment in new turbine production facilities. Table 5 provides an overview of selected federal programs affecting the U.S. wind power industry.

Table 5. Selected Energy Programs Affecting the U.S. Wind Industry

Program	Expiration Deadlines for Wind Generation/Manufacturing Projects
Production Tax Credit[a]	December 31, 2013 (begin construction)[d]
Investment Tax Credit[b]	December 31, 2013 (begin construction)
Advanced Manufacturing Tax Credit	Capped at $2.3 billion; 100% Allocated
1603 Cash Grant in Lieu of Tax Credit[c]	December 31, 2011 (begin construction) December 31, 2012 (placed in service)
1705 Loan Guarantee Program	September 30, 2011 (commence construction)
Bonus Depreciation Schedule	December 31, 2011, for 100% first-year bonus depreciation December 31, 2013, for 50% bonus

Source: Wind Energy Manufacturers Association, Supply Chain Issues from Tier 1 Suppliers and Component Makers, http://www.slideshare.net/LeslieFeen/supply-chain-issues-from-tier-1-suppliers-and-component-makerswind-power-manufacturing-amp-supply-cha.

[a] The PTC had expired on December 31, 2012, but was extended by Congress until the end of 2013.

[b] The taxpayer who presumably is the owner of the relevant wind power project prior to December 31, 2013, can opt for a 30% ITC in lieu of the PTC. The ITC allows offshore wind facilities to receive 30% of capital expenditures for projects entering construction by December 31, 2013. The 30% ITC for small commercial wind energy property extends through December 31, 2016.

[c] The taxpayer owning the relevant wind power project can opt for a 30% cash grant from the U.S. Department of Treasury instead of a PTC, and then select a one-time cash grant instead of tax credits. The American Taxpayer Relief Act did not modify or extend the 1603 grant program.

[d] The extension of the PTC included a new provision allowing wind projects that begin construction in 2013 to qualify, whereas previously the tax credit was available only to wind power generation projects that were placed in service and were operating by the end of the calendar year in which the credit was set to expire.

Production Tax Credit (PTC)/Investment Tax Credit (ITC)

The PTC, the main federal policy tool in the deployment of U.S. wind power, was first adopted during the Administration of President George H. W. Bush as part of the Energy Policy Act of 1992 (P.L. 102-486). It has been a significant driver of the recent growth of the U.S. wind industry, but it is not a permanent part of the tax code and has lapsed on a number of occasions. In each of the years during which the PTC completely lapsed (1999, 2001, and 2003), meaning that it expired prior to being renewed, the level of additional deployed wind capacity slowed or collapsed when compared to the previous year's total: 93% in 2000, 73% in 2002, and 77% in 2004.[116] Yet, when the PTC incentive was extended in 2000, 2002, 2004, 2005, 2007, 2008, 2009, and 2013, the industry responded positively, increasing wind power capacity compared to the previous year. 2010 was an exception to this trend with a drop in wind capacity of nearly 50% from 2009, even with the PTC in place.[117] In 2011, at 6,816 MW, annual installed wind capacity increased by 30% over the previous year. The annual cost of the PTC is estimated at about $1 billion a year.[118] Most recently, Congress provided a one-year extension of the PTC for wind through December 31, 2013, as part of the American Taxpayer Relief Act of 2012 (ATRA; P.L. 112-240).[119] The PTC provides a per kilowatt-hour (kWh) income tax benefit over the first 10 years of a wind project's operations, which is 2.2 cents per kWh, and is a critical factor in financing new wind farms. In order to qualify, a wind farm must start construction while the credit is in place, which would be by the end of 2013.[120] Wind project developers may elect to receive a 30% investment tax credit (IRC §48) in place of the PTC if the projects begin construction prior to the end of 2013.[121]

AWEA continues to advocate for a phase out of the PTC over six years, which it argues would encourage continued investment in the industry and would allow for extended growth of domestic turbine manufacturing.[122] The Governors' Wind Energy Coalition has called for a multi-year extension of the PTC of at least four years.[123]

Advanced Energy Manufacturing Tax Credit (MTC)

The Advanced Energy Manufacturing Tax Credit, also referred to as Section 48C of the Internal Revenue Code, was authorized in Section 1302 of the American Recovery and Reinvestment Act.[124] The MTC provided a 30% credit for companies for investments in new, expanded, or reequipped clean energy domestic manufacturing facilities built in the United States. Wind, solar panels, and electric vehicle batteries were among the 183 projects funded through the MTC before reaching its cap of $2.3 billion in 2010. The Obama Administration requested another $5 billion for the 48C tax program. An extension of the MTC has been proposed through the Security in Energy and Manufacturing Act of 2011 (S. 591), or SEAM Act. It includes one significant change from the original MTC; higher priority would be given to facilities that manufacture—rather than assemble—goods and components in the United States.[125] Fifty-two wind manufacturing projects were awarded $364 million in tax credits under the MTC program.[126] Beneficiaries included many manufacturers that were already active, or that had announced that they intend to open new facilities, in the United States. Selected manufacturers of wind turbines, blades, towers, and gears that received tax credits under the 48C program are listed in Appendix C.

Other Wind-Related Programs

Tax benefits for wind projects include accelerated tax depreciation and bonus depreciation; the latter allowed wind farm owners to write off more than 50% of the capital costs of building a wind farm in 2008, 2009, and 2010. The 2010 Tax Act[127] increased the first-year bonus depreciation to 100% for new qualified property acquired and placed in service between September 8, 2010, and December 31, 2011, rather than 50% for the qualifying property. Bonus depreciation dropped to the lower 50% rate in 2012.[128] The American Taxpayer Relief Act provided a one-year extension to the end of 2013. Another ARRA incentive is a grant system administered by the U.S. Treasury Department. In lieu of tax credits, wind projects can receive a cash payment of up to 30% of the qualified capital costs. The Section 1603 Treasury cash grant program allows developers to opt for a cash payment instead of a tax break. To qualify, construction had to begin by December 31, 2011.[129] Wind projects under construction by year-end 2011 must be placed in service by December 31, 2012. Many in the wind industry credited the grants for keeping the sector healthy during the 2008 and 2009 recession.[130] The Section 1705 loan program, a temporary ARRA program administered by the Department of Energy, authorized loan guarantees for certain renewable energy projects, including wind projects. The program, which funded 26 projects, including four wind generation projects, expired on September 30, 2011. The combined wind commitments totaled $1.7 billion, or 9% of the $18.8 billion in 1705 program funding.[131] The Caithness Shepherds Flat wind generation project, said to be one of the largest onshore wind farms in the world, received a $1.3 billion loan.[132] GE manufactured the wind turbines. Loan guarantees were also extended to three other wind generation projects: Kahuku Wind Power, Granite Reliable, and Record Hill Wind. No wind turbine manufacturers were funded under the 1705 program.[133]

State Renewable Portfolio Standards

State renewable portfolio standards have encouraged the growth of the U.S. wind energy industry by requiring companies that sell electricity to retail customers to obtain a specified share of their electricity from renewable generation.[134] As of June 2012, mandatory RPS programs existed in 29 states and the District of Columbia.[135] The U.S. wind industry has long called for a national standard to increase investor confidence in the sector's long-term prospects. No such measure has passed Congress, although bills to establish national renewable standards have been passed by the Senate on three occasions and by the House of Representatives once.[136]

CONCLUSION

The expansion of the U.S. wind power manufacturing base will depend, at least in part, on government policy decisions. The production costs of U.S. plants that make turbine components appear to be competitive with those in other countries, and the difficulty and expense of transporting very bulky products over long distances serves as an obstacle to import competition.

Nonetheless, there are several obstacles that may impede the expansion of wind energy manufacturing in the United States. One is the history of policy-induced boom-and-bust cycles in wind energy investment, which may lead wind turbine manufacturers and component suppliers to conclude that future U.S. demand for their products is too uncertain. Another significant challenge affecting the sector's future is the availability of adequate transmission for power generated by wind farms. Most wind farms are located at a distance from the urban areas where most electricity is consumed, and a shortage of transmission capacity could hamper wind farm creation or expansion. Congress may wish to evaluate the seriousness of transmission issues in the context of other federal efforts to support wind generation.[137] The structure of the wind manufacturing industry is also likely to undergo significant change. As is typical in budding industries, a large number of companies now compete in wind manufacturing. Mergers and failures are likely to lead to consolidation as the sector matures. As this report describes, competition in the wind turbine sector from new Asian entrants will likely become more significant in future years, but it is still unclear whether many of these companies have the technological abilities and financial resources to become significant players in the U.S. market.

APPENDIX A. GLOBAL WIND TURBINE MANUFACTURERS

Table A-1. Global Wind Turbine Manufacturers by Original Equipment Manufacturers (OEMs) Top 10 by Annual Market Share (installed capacity), 2009, 2010, 2011

Manufacturer	Location of Headquarters	2009	Manufacturer	Location of Headquarters	2010	Manufacturer	Location of Headquarters	2011
Vestas	Denmark	12.50%	Vestas	Denmark	14.80%	Vestas	Denmark	12.90%
GE	U.S.	12.4	Sinovel	China	11.1	Goldwind	China	9.4
Sinovel	China	9.2	GE	U.S.	9.6	GE	U.S.	8.8
Enercon	Germany	8.5	Goldwind	China	9.5	Gamesa	Spain	8.2
Goldwind	China	7.2	Enercon	Germany	7.2	Enercon	Germany	7.9
Gamesa	Spain	6.7	Suzlon	India	6.9	Suzlon	India	7.7
Dongfang	China	6.5	Dongfang	China	6.7	Sinovel	China	7.3
Suzlon	India	6.4	Gamesa	Spain	6.6	Guodian United Power	China	7.1
Siemens	Germany	5.9	Siemens	Germany	5.9	Siemens	Germany	6.3
Repower	Germany	3.4	United Power	China	4.2	MingYang WindPower	China	2.9

Sources: Vestas Remains Top Wind Turbine Maker, Goldwind Is Second, BTM Consult, March 26, 2012.

Notes: Market share data is reported in MW terms and is based on installations in the year in question, not on turbine shipments or orders.

APPENDIX B. SELECTED EXAMPLES OF U.S. WIND TURBINE PRODUCTION FACILITIES

Table B-1. Examples: U.S. Turbine Production Facilities

Wind Turbine Manufacturer/Headquarters	U.S. Location	Wind Turbine Production Facilities
Gamesa (Spain)	PA	Gamesa, the first foreign-based wind turbine manufacturer to set up full production facilities in the United States, opened a plant at a former U.S. Steel factory in Ebensburg, PA, in 2005. Gamesa also operates a nacelle manufacturing plant in Fairless Hills, PA. It invested over $175 million in these plants and received $15 million in state subsidies and tax credits. In 2012, Gamesa reduced its workforce at the two plants by 160 workers, citing lower demand and the possible expiration of the production tax credit.a
Suzlon (India)	MN	Suzlon opened a rotor blade manufacturing facility in Pipestone, MN, in 2006, with an investment of $8.5 million, its first manufacturing facility outside India. That factory, which once employed over 500 workers, was idled in 2010 save for blade repair and customer service operations, and most of its workers have been laid off.b Suzlon acquired the German manufacturer REpower in March 2012.
Siemens (Germany)	IA/KN	Siemens opened a wind turbine blade factory in Fort Madison, IA in 2007 and a nacelle factory in Hutchinson, KS in 2010. In 2012, Siemens announced a reduction of its U.S. wind power workforce by more than 37%, citing low natural-gas prices, a slow economic recovery, and the possible expiration of the PTC.c
Vestas (Denmark)	CO	Vestas opened a blade production plant in Windsor, CO, and an R&D center in Louisville, CO in 2010 and another blade manufacturing plant in Brighton, CO in 2012. Vestas received an incentive package of approximately $4 million to invest in Colorado, including grants, tax rebates, and job-training funds. In 2012, employment at Vestas's factories in Colorado fell to 1,200 from more than 1,700 at the start of the year.d

Wind Turbine Manufacturer/Headquarters	U.S. Location	Wind Turbine Production Facilities
Nordex (Germany)	AR	Nordex opened a nacelle assembly plant in 2010, which represented a $40 million investment. It expected to open a blade manufacturing plant in 2012 in Jonesboro, AR, but those plans have yet to be realized.
Mitsubishi (Japan)	AR	Mitsubishi in 2012 announced that it would indefinitely delay the opening of its nacelle assembly facility in Fort Smith, AR. The company had stated that the facility would employ 400 workers.e

Source: Compiled by CRS from various sources including company annual reports, press releases, news reports, and information from AWEA.

[a] Business News in Brief, July 6, 2012, Philadelphia Inquirer, http://articles.philly.com/2012-07-06/business/32566874_1_wind-turbines-reports-passenger-refinery.

[b] David Shaffer, *Pipestone Wind-Turbine Factory Idled; 110 Layoffs*, StarTribune, November 1, 2010, http://www.startribune.com /business/106490454.html?refer=y.

[c] Laura DiMugno, *More PTC Fallout: Siemens Cuts 37% of its U.S. Wind Energy Workforce*, North American WindPower, September 18, 2012, http://nawindpower.com/ e107_plugins/content/content.php?content.10418#.UMZJnXe3qVk.

[d] *Vestas Statement on Colorado Layoffs*, Denver Business Journal, October 11, 2012, http://www.bizjournals.com/denver/news /2012/10/11/vestas-statement-on-coloradolayoffs.html?page=all.

[e] Rusty Garrett, *Mitsubishi Mothballs Fort Smith Wind Turbine Plant*, Times Record, April 2, 2012, http://swtimes.com/sections /news/mitsubishi%E2%80%9Cmothballs%E2%80%9D-fort-smith-wind-turbine-plant.html.

APPENDIX C. 48C MANUFACTURING TAX CREDIT

Table C-1. Selected Wind Manufacturers Receiving Section 48C Manufacturing Tax Credit

Applicant	Tax Credit Requested	State
Siemens	$28,328,379	IL
Nordex	$22,153,500	AR
Merrill Technologies Group	$22,021,500	MI
Vestas	$21,600,000	CO
Vestas	$21,589,200	CO
Tindall Corporation	$16,750,500	SC
Winergy	$12,786,000	IL
Brevini	$12,750,000	IN
Vela Gear Systems	$11,604,440	MI
Vestas	$8,580,600	CO
Hexcel Corporation	$8,139,510	CO
TPI Composites	$5,135,241	NE
Mitsubishi Power Systems	$5,100,000	AR
Siemens	$4,331,700	KS
TPI Composites	$3,902,921	IA
Siemens	$3,450,900	IA
Nordic Windpower	$3,000,000	ID
Alstom	$2,725,800	TX

Source: Strategic Partnerships, Inc., http://www.spartnerships.com/reports/ARRA%20Energy%20Manufacturing %20Tax%20 Credit%20Awards.pdf.

Notes: A tax credit is a "dollar for dollar" reduction in tax liability. As an example, if a manufacturer earns $10 million and owes $3.5 million in taxes, then a $1million tax credit would reduce the company's tax liability from $3.5 million to $2.5 million.

End Notes

[1] Employment data for the U.S. wind energy sector is currently only reported by the American Wind Energy Association (AWEA). Recent statistics can be found in AWEA's annual report, *U.S. Wind Industry Annual Market Report Year Ending 2011*, p. 49.

[2] BLS does not publish earnings data specific to the wind power industry, but it estimates that earnings for engineers in wind power are comparable to earnings for engineers in general. James Hamilton and Drew Liming, *Careers in Wind Energy*, Bureau of Labor Statistics, September 2010, pp. 10-11, http://www.bls.gov/green/wind_energy/ wind_energy.pdf.

[3] The U.S. Energy Information Administration (EIA) reports wind energy represented about 4% of U.S. power generating capacity, and 3% of total U.S. electricity generation in 2011.

[4] EIA, *Direct Federal Financial Interventions and Subsidies in Energy in Fiscal Year 2010*, July 2011, http://docs.wind-watch.org/US-subsidy-2010.pdf.

[5] For a detailed discussion on energy tax incentives see CRS Report R41953, *Energy Tax Incentives: Measuring Value Across Different Types of Energy Resources*, by Molly F. Sherlock.

[6] EIA, Renewable and Alternative Fuels, *Renewable Portfolio Standards and State Mandates by State*, August 2011, http://www.eia.gov/cneaf/solar.renewables/page/trends/table28.html.

[7] The Clean Energy Standard Framework announced by the White House in 2011 is discussed in CRS Report, R41720, *Clean Energy Standard: Design Elements, State Baseline Compliance and Policy Considerations*, by Phillip Brown.

[8] AWEA, U.S. Wind Industry Annual Market Report 2011, p. 40.

[9] Worldwatch Institute, *Made in China, or Made by China? Chinese Wind Turbine Manufacturers Struggle to Enter Own Market*, http://www.worldwatch.org/node/3931.

[10] U.S. Department of Energy, *A National Offshore Wind Strategy: Creating an Offshore Wind Energy Industry in the United States*, February 2011, http://www1.eere.energy.gov /windandhydro/pdfs/national_offshore_wind_strategy.pdf.

[11] Cape Wind, "Cape Wind Completes Permitting Process," press release, January 7, 2011, http://www.capewind.org/ news1174.htm.

[12] AWEA, U.S. Wind Industry Annual Market Report 2011, p. 57.

[13] Each of the seven projects is eligible to receive up to $4 million for the engineering, site evaluation, and planning phase of its project. Upon completion of phase one, up to three may receive a limit of $47 million over four years, subject to congressional appropriations, for construction and installation. Department of Energy, "Energy Department Announces New Investments in Pioneering U.S. Offshore Wind Projects," press release, December 12, 2012, http://energy.gov/articles/energy-department-announces-new-investments-pioneering-us-offshore-wind-projects.

[14] AWEA, *2011 U.S. Small Wind Turbine Market Report*, June 2012, p. 22, http://www.awea.org /learnabout/smallwind/upload/AWEA_SmallWindReport-YE-2011.pdf.

[15] AWEA, *2010 Small Wind Turbine Global Market Study*, Year Ending 2009, p. 18, http://www.awea.org/learnabout/ smallwind/upload/2010_AWEA_Small_Wind_Turbine_Global_Market_Study.pdf.

[16] Charles F. Brush, an American inventor, constructed the first modern wind turbine in 1888, in Cleveland, OH, for the purpose of electricity generation. He used it to power his home. Thereafter, other Americans such as Palmer C. Putman built wind turbine generators, mostly for farm use, at a time when electricity distribution systems had not yet been installed. U.S. manufacturers of early wind turbine generators included Jacobs Wind and Parris-Dunn. The rural electrification project of 1936 effectively killed the wind-generated

power market in the United States until the early 1970s. For more information, see Windsector, *The First Wind Turbine in the United States*, April 17, 2011, http://windsector.tumblr.com/post/4711554356/the-first-wind-turbine-in-america.

[17] Janet Swain, "The Role of Government in the Development and Diffusion of Renewable Energy Technologies: Wind Power in the United States, California, Denmark, and Germany," (Ph.D. dissertation, Fletcher School of Law and Diplomacy, 2001), pp. 200-203. This dissertation notes that by 1991 77% of the world's wind capacity was installed in California.

[18] Geoffrey Jones and Loubna Bouamane, "Historical Trajectories and Corporate Competences in Wind Energy," (Working Paper 11-112, Harvard Business School, 2011), p. 32. http://www.hbs.edu/research/pdf/11-112.pdf.

[19] Over 15,000 medium-sized wind turbines were installed in California between 1981 and 1986. See Union of Concerned Scientists, Briefing on How Wind Energy Works, http://www.ucsusa.org/clean_energy/coalvswind/ brief_wind.html.

[20] Large wind turbines are often called utility-scale because they generate enough power for utilities, or electric companies, to sell.

[21] Jens Vestergaard, Lotte Brandstrup, and Robert Goddard, *Industry Formation and State Intervention: The Case of the Wind Turbine Industry in Denmark and the United States*, Published in the Academy of International Business Conference Proceedings, p. 16-18, http://pure.au.dk/portal/files/2552/windmill_paper2.pdf.

[22] Lester Brown, *World on the Edge: How to Prevent Environmental and Economic Collapse*, Earth Policy Institute, Supporting Data Showing Cumulative Installed Wind Power Capacity and Annual Additions to the United States, 1980-2009, 2011, http://www.earth-policy.org/books/wote/wote_data_topic.

[23] An overview of policy instruments used by various governments to promote renewables, including wind power, can be found on the Renewable Energy Policy Network website at http://www.ren21.net/RenewablesPolicy/ PolicyInstruments/tabid/5608/Default.aspx.

[24] The wind turbine industry advanced in Europe, specifically in Denmark, beginning in the early 20th century based largely on the wind turbines constructed by Poul la Cour. For background, see Jens Vestergaard, Lotte Brandstrup, and Robert Goddard, *"A Brief History of the Wind Turbine Industries in Denmark and the United States,"* (Academy of International Business, 2004), http://www.hha.dk/man/cmsdocs/publications /windmill_ paper1.pdf.

[25] AWEA, *U.S. Wind Industry Annual Market Report 2011*, p. 4. Utility-scale wind turbines as defined by AWEA are large turbines with generating capacity of 100 kW and larger.

[26] China faces major challenges with grid connection of installed wind turbines, as some projects in China have to wait several months before being connected to the national grid. Thus, the United States continues to exceed China in grid connected wind power capacity. China issues two figures when it reports its wind power data. By year-end 2011, China reported that it installed 62.4 gigawatts (GW) of onshore wind power, but only 45 GW was operational and connected to the grid. In other markets, it is common practice to count all turbines as soon as they are grid connected and producing electricity. For more information see REN21, *Renewables 2011 Global Status Report*, Table R8, p. 104, http://www.ren21.net/Portals/97/documents/GSR/GSR2012_low%20res_FINAL.pdf.

[27] Global Wind Energy Council (GWEC) *Global Wind 2011 Report*, March 2012, p. 11, http://www.gwec.net/fileadmin/documents/Publications/Global_Wind_2007_report/ GWEC%20Global%20Wind%20Report%202010%20low%20res.pdf.

[28] Ryan Wiser and Mark Bolinger, *2011 Wind Technologies Market Report*, Lawrence Berkeley National Laboratory, August 2012, p. 19, http://www1.eere.energy.gov/wind /pdfs /2011_wind_technologies_market_report.pdf.

[29] Independent power producers are companies that produce power that they sell to electric utilities.

[30] A list of some 975 wind farms in the United States can be accessed at Windpower's wind turbine and wind farms database, http://www.thewindpower.net/country-datasheet-windfarms-4-usa.php.

[31] Ekopolitan, "World Turbine Market Shares, 2008-2011, Installed Capacity" BTM Estimates, http://www.ekopolitan.com/tech/global-wind-turbine-market-shares.

[32] Zond was purchased by Enron Wind in 1997. In 2002, GE, which had long produced turbines for power generation, acquired Enron Wind's fully integrated wind power capacity including its line of wind turbine generators.

[33] Clipper Windpower does not rank among the top 10 global wind turbine manufacturers and it has found itself squeezed in the United States, its main market, by larger competitors. In December 2010, United Technologies Corporation purchased Clipper, which it sold in August 2012 to a private equity firm, Platinum Equity. Clipper has downsized its operations and reduced its staff to fewer than 100 employees.

[34] GWEC reports China's wind market doubled every year between 2006 and 2009, and it has been the largest annual market by installed capacity in the world since 2009. GWEC, *Global Wind 2011 Report*, March 2012, p. 12, http://gwec.net/wp-content/uploads/2012/06 /Annual_report_2011_lowres.pdf.

[35] Joshua Meltzer, *The United States and China: The Next Five Years*, The Brookings Institution, May 19, 2011, p. 17, http://www.brookings.edu/~/media/Files/events /2011/0519_us_ china/20110519_us_china_panel4.pdf.

[36] Geoffrey Jones and Loubna Bouamane, *Historical Trajectories and Corporate Competences in Wind Energy* (Working Paper 11-112, Harvard Business School, 2011), p. 55.

[37] Joanna Lewis, *Why China is acting on Clean Energy: Successes, Challenges, and Implications*, Georgetown University, October 12, 2012, p. 12, http://files.eesi.org/Lewis101212.pdf.

[38] John McDonald, *Wind Power Market Opportunity Profile, China*, British Columbia Trade and Investment, 2009, pp. 2-3, https://trade.britishcolumbia.ca/Export/Markets /Documents /China_WindPower.pdf.

[39] Joanna Lewis, *Can Green Sunrise Industries Lead the Drive into Recovery? The Case of the Wind Power Industry in China and India*, United Nations Industrial Development Organization, 2010, p. 7, http://www.unido.org/fileadmin/ user_media/Publications /RSF_DPR/WP202009_Ebook.pdf.

[40] Joanna I. Lewis, *Building a National Wind Turbine Industry: Experiences from China, India, and South Korea*, Georgetown University, Int. J. Technology and Globalization, Vol. 5, Nos. 3/4, 2011, pp. 290-293, http://www.china.tu-berlin.de/fileadmin/fg57/SS_2012 /Umwelt/Lewis_windenery.pdf. For more information on the major players in the South Korean wind industry, see a report by the Maine International Trade Center, *"Opportunities for Maine Companies in Korean New and Renewable Energy (NRE) Markets*," pp. 7-9, November 2010. http://www.mitc.com/PDFs/RenewableEnergyinKorea_Report.pdf.

[41] AWEA, *U.S. Wind Industry Annual Market Report 2011*, p. 7.

[42] Ryan Wiser and Mark Bolinger, *2011 Wind Technologies Market Report*, Lawrence Berkeley National Laboratory, August 2012, p. 15, http://www1.eere.energy.gov/wind/pdfs /2011_wind_technologies_market_report.pdf.

[43] AWEA's 3rd Quarter 2012 Market Report found that total installations through the first nine months of 2012 reached 4,728 MW, compared to 3,370 for the same period in 2011.

[44] AWEA, *Anatomy of a Wind Turbine*, http://www.awea.org/issues/supply_chain/Anatomy-of-a-Wind-Turbine.cfm.

[45] A detailed description of the components in a wind turbine can be found in *Wind Turbine Development: Location of Manufacturing Activity*, by George Sterzinger and Matt Svrcek, Renewable Energy Policy Project, September 2004. http://www.repp.org/articles/static/1/binaries/WindLocator.pdf.

[46] The German manufacturer Enercon has built the world's largest turbine model to date, the Enercon E-126, which can generate up to 7 MW of power.

[47] Economic and Workforce Development Program California Community Colleges, *Wind Turbine Technicians*, September 2009, p. 35, http://www.coeccc.net/Environmental_Scans/wind_scan_sw_09.pdf.

[48] Chi-Jen Yang, Eric Williams, and Jonas Monast, *Wind Power: Barriers and Policy Solutions*, Nicholas School of the Environment at Duke University, November 2008, pp. 14-15. http://nicholasinstitute.duke.edu/climate/electricity/windpower-barriers-and-policy-solutions.

[49] Jacob Funk Kirkegaard, Thilo Hanemann, and Lutz Weischer, *It Should Be a Breeze: Harnessing the Potential of Open Trade and Investment Flows in the Wind Energy Industry*, Peterson Institute for International Economics, December 2010, p. 41, http://www.iie.com/publications/wp/wp09-14.pdf.

[50] David Wilburn, *Wind Energy in the United States and Materials Required for the Land-Based Wind Turbine Industry From 2010 through 2030*, U.S. Geological Survey, Scientific Investigations Report 2011-5036, 2011, pp. 7-8, http://pubs.usgs.gov/sir/2011/5036/sir2011-5036.pdf.

[51] Gloria Ayee, Marcy Lowe, and Gary Gereffi, et al., *Manufacturing Climate Solutions Carbon Reducing Technologies and U.S. Jobs*, Center on Globalization Governance and Competitiveness, Wind Power, September 22, 2009, p. 10, http://www.cggc.duke.edu/environment/climatesolutions/greeneconomy_Ch11_WindPower.pdf.

[52] BlueGreen Alliance, *Clean Energy Economy Report 2009*, June 15, 2009, p. 3, http://www.repp.org/articles/ BGA_Repp.pdf.

[53] U.S. Department of Energy, *2011 Wind Technologies Market Report*, August 2012, p. 33.

[54] One analysis of vertical integration among wind OEMs indicates that Suzlon and Enercon have significant in-house production and high or very high levels of vertical integration; Siemens and Vestas fall in the middle; and GE is less vertically integrated than many other manufacturers, relying on outside suppliers for blades, gearboxes, generators, castings and forgings, and towers. Josh Lutton, *Wind Turbine Manufacturer Recommendations (Round 2)*, Woodlawn Associates, April 27, 2010, p. 6, http://www.woodlawnassociates.com/uploads/Woodlawn_Associates_WT_Recs_-_R2_100427.pdf.

[55] Dan Ancona and Jim McVeigh, *Wind Turbine—Materials and Manufacturing Fact Sheet*, Office of Industrial Technologies, U.S. Department of Energy. August 29, 2001. http://www.perihq.com/documents/WindTurbine-MaterialsandManufacturing_FactSheet.pdf.

[56] Michelle Avis and Preben Maegaard, *Worldwide Wind Turbine Market and Manufacturing Trends*, Xmire, January 2008, p. 21, http://www.folkecenter.net/mediafiles/folkecenter/pdf/Market_and_Manufacturer_Trends.pdf.

[57] Manufacturers like Suzlon have experienced recent failures of their turbines. Reliability and performance are critical factors affecting shareholder value, the reputation, and future growth of any wind OEM.

[58] AWEA provided these statistics to CRS via email on December 5, 2012.

[59] AWEA, *Policy and Manufacturing: Demand-Side Policies Will Fuel Growth in the Wind Manufacturing Sector*, 2011, p. 3, http://www.thenewnorth.com/resources /mwgpolicy paper.pdf.

[60] Andrew David, *Impact of Wind Energy Installations on Domestic Manufacturing and Trade*, U.S. International Trade Commission, July 2010, p. 7, http://www.usitc.gov/publications /332/working_papers/ID-25.pdf.

[61] Transporting wind turbines, which requires special trucks, railroad carriages, and cranes, is difficult because of their unusual weight, length, and shape. For example, a typical nacelle weighs between 50 and 70 tons. Blades can run from 110 feet to 145 feet. Towers can weigh 70 tons. According to some estimates, transportation costs can account for up to 20% of the installed cost of a wind turbine. Estimates from AWEA suggest that per-turbine transportation and logistics costs range from $100,000 to $150,000. For more information see, *The Logistics of Transporting Wind Turbines: Reducing Inefficiencies, Costs, and Community Impact by Streamlining the Supply Chain*, CN White Paper 2009, http://www.cn.ca/documents/WhitePapers/Transporting-Wind-Turbines-White-Paper-en.pdf.

[62] LM Wind Power, LM Wind Power Group—Facts, http://www.lmwindpower.com/upload/ lmwp_factsheet_groupuk_020511.pdf.

[63] AWEA, *U.S. Wind Industry Annual Market Report, 2011*, 2012, p. 38.

[64] The American Recovery and Reinvestment Act (ARRA) of 2009 included grants, loans, and tax credits.

[65] Chisaki Watanabe, "Mitsubishi Heavy to Suspend U.S. Wind Factory on Sluggish Demand," April 2, 2012.

[66] Suzlon, which once employed more than 500 U.S. workers who made rotor blades used in wind turbines in Pipestone, MN, now has about 30 employees there, engaged primarily in blade repair work and customer service. Debra Fitzgerald, "Suzlon Shifts Focus of Pipestone Plant," *Pipestone County StarOnline*, February 15, 2012.

[67] FAG, a unit of the Schaeffler Group, has produced bearings for wind turbines for over 30 years. It has a U.S. factory in Joplin, MO. See FAG, *"Expertise in Bearing Technology and Service for Wind Turbines,"* March 2010, http://www.schaeffler.com/remotemedien/media /_shared_media/library/schaeffler_2/brochure/downloads_1/ pwe_de_en.pdf.

[68] Timken, headquartered in Ohio, is a global supplier of bearings with a full line for the wind industry. Timken produces ultra-large bore bearings for wind turbines in South Carolina and also operates wind-bearing production manufacturing facilities in Brazil, China, and Romania. It also provides clean steel for wind energy from its facilities in Canton, OH. See http://www.timken.com/en-us/about/Pages/Locations.aspx.

[69] "Siemens & Winergy Open Wind Turbine Manufacturing Plant," *Renewable Energy World.com*, August 31, 2009. http://www.renewableenergyworld.com/rea/news /article/2009/08/siemens-winergy-open-turbine-manufacturing-plant.

[70] Winergy, Production Locations, http://www.winergy-group.com/cms/website.php?id=/en /about-winergy/ locations.htm.

[71] Ryan Wiser and Mark Bolinger, *2011 Wind Technologies Market Report*, National Renewable Energy Laboratory, August 2012, p. 16, http://www1.eere.energy.gov/wind /pdfs/2011_ wind_technologies_market_report.pdf.

[72] AWEA reports at least 30 facilities in Michigan, more than 50 companies in Ohio, and about 15 factories in Pennsylvania now manufacture components for the wind industry. AWEA, *State-specific wind energy fact sheets, updated through the 3rd Quarter 2012*, http://www.awea.org/learnabout/publications/factsheets/factsheets_state.cfm.

[73] AWEA employment data were provided to CRS via email on December 4, 2012, and are based on surveys and modeling.

[74] AWEA, *U.S. Wind Industry Annual Market Report, 2011*, 2011, p. 45. AWEA is the only source of nationwide employment statistics, as the U.S. government does not currently track employment in the wind industry. Measurement of employment in wind turbine manufacturing is complicated by the fact that no single industry codes exist to isolate wind power establishments or wind turbine and wind components establishments. The North American Industry Classification System (NAICS) places wind turbine manufacturers within the Turbine and Turbine Generator Set Units manufacturing industry (NAICS 333611), which comprises "establishments primarily engaged in manufacturing turbines (except aircraft) and complete turbine generator set units, such as steam, hydraulic, gas, and wind." BLS reports 29,070 total jobs in this industry in 2011, with employment increasing every year since 2005, when it had 19,500 employees.

[75] U.S. Department of Energy, *20% Wind Energy by 2030: Increasing Wind Energy's Contribution to U.S. Electricity Supply*, July 2008, p. 207, http://www.20percentwind.org /20percent_wind_energy_report_revOct08.pdf. DOE estimates are based on major component assumptions that by 2030 80% of blades, 50% of towers, and 42% of turbines installed in the United States would be manufactured domestically.

[76] AWEA, *Policy and Manufacturing: Demand-Side Policies Will Fuel Growth in the Wind Manufacturing Sector*, 2011, p. 9, http://www.thenewnorth.com/resources /mwgpolicy paper.pdf.

[77] Gerald Susman and Amy Glasmeier, "*Industry Structure and Company Strategies of Major Domestic and Foreign Wind and Solar Energy Manufacturers: Opportunities for Supply Chain Development in Appalachia*," Smeal College of Business, November 20, 2009, p. 38, http://www.arc.gov/assets/research_reports/WindandSolarEnergy.pdf.

[78] Precisely tracking trade flows in the wind industry is complicated because the standard Harmonized Commodity Coding and Classification System (HS) does not have separate harmonized trade categories for all wind turbines and their components. Wind turbines and components are classified under several HS codes. Wind-powered generating sets (HS 8502.31) is the main category, which includes fully assembled wind turbines, but may also cover components such as blades and hubs when they are imported with the nacelle. However, when imported separately other individual turbine components (e.g., generators (HS 8501.64), towers (7308.20), and blades and other components (8412.90 and 8503.00) may be traded under other HS headings. Importantly, goods that are not used in wind turbines are also included in these categories. But, the ITC reports, wind accounts for a significant portion of trade in each dual use category and appears to be a major driver of import growth in those HS headings. For a complete discussion see, USITC, "*Wind Turbines: Industry and Trade Summary*," by Andrew David, June 2009.

[79] Andrew David, *Shifts in U.S. Wind Turbine Equipment Trade in 2010*, U.S. International Trade Commission, USITC Executive Briefing on Trade, June 2011, http://www.usitc.gov /publications/332/executive_briefings/ wind_EBOT_commission_review_final2.pdf.

[80] Dr. Rimtalg Lee, *Status and Forecast of Wind Energy In Korea*, San Francisco, CA, March 2, 2009, pp. 7-9, http://www.asiapacificpartnership.org/pdf/PGTTF/wind-event/March_2 /StatusnForecastofWindEnrgyKorea.pdf.

[81] Ginger Gardiner, "Windpower 2012 Report," *Composites Technology*, August 1, 2012, http://www.compositesworld.com/articles/windpower-2012-report. BTM Consult reports total capacity of Chinese wind turbine exports increased to 222 MW in 2011 from almost 17 MW in 2008.

[82] Vestas, Company Structure, Vestas China, http://www.vestas.com/en/about-vestas/company-structure/vestaschina.aspx.

[83] United Steelworkers, *United Steelworkers' Section 301 Petition Demonstrates China's Green Technology Practices Violate WTO Rules*, http://assets.usw.org/releases/misc/section-301.pdf.

[84] China's Special Fund for Wind Power Equipment Manufacturing provided individual grants ranging from $6.7 million to $22.5 million to Chinese wind turbine manufacturers in exchange for using domestic parts and components instead of imported ones. For more information on China's Special Fund see USTR's June 7, 2011 press release, "China Ends Wind Power Equipment Subsidies Challenged by the United States in WTO Dispute," http://insidetrade.com/iwpfile.html?file=jun2011%2Fwto2011_1868a.pdf.

[85] The Wind Tower Trade Coalition comprises Broadwind Towers of Manitowoc, WI; DMI Industries of Fargo, ND; Katana Summit LLC of Columbus, NE; and, Trinity Structural Towers of Dallas, TX.

[86] In the countervailing duty case, DOC found that Chinese wind tower manufacturers, including CS Wind and Titan Wind, benefited from Chinese subsidy programs. See *Commerce Preliminarily Finds Countervailable Subsidization of Imports of Utility Scale Wind Towers from the People's Republic of China (China)*, May 30, 2012, http://ia.ita.doc.gov/ download/factsheets/factsheet-prc-towers-cvd-prelim-20120530.pdf.

[87] The DOC preliminarily assessed duties include 30.93% on Chengxi Shipyard, 20.85% on Titan Wind Energy, 26.25% on CS Wind Corporation, Guodian United Power Technology Baoding, and Sinovel, and a China-wide rate of 72.69%. For a DOC anti-dumping fact sheet, see U.S. Department of Commerce, International Trade Administration, *Commerce Preliminarily Finds Dumping of Imports of Utility Scale Wind Towers*, July 27, 2012, http://ia.ita.doc.gov/ download/factsheets/factsheet-china-vietnam-uswt-ad-prelim-20120 727.pdf.

[88] The Harmonized Tariff Schedule classifies wind towers under towers and lattice masts (HS 7308.20). Not all the towers in this category are wind towers.

[89] Wind blades are classified under the tariff lines for parts of other engines and motor (HS 8412.90) and parts of generators (HS 8503.00). Not all shipments in this category are wind-related.

[90] Precisely how many wind turbine components are made in the United States and how many are imported is a debatable issue. U.S. content need only be disclosed on a few products, namely automobiles, textiles, wool, and fur products. For most other products, no law requires disclosure of domestic content. In the case of automobiles, the American Automobile Labeling Act (AALA) requires automobile assemblers to include labels that specify the percentage value of the U.S./Canadian parts content of each vehicle sold in the United States.

[91] Ryan Wiser and Mark Bolinger, *2011 Wind Technologies Market Report*, Lawrence Berkeley National Laboratory, August 2012, p. 23, http://www1.eere.energy.gov/wind/pdfs /2011_ wind_technologies_market_report.pdf.

[92] Michael A. Peck, *Briefing by Michael A. Peck to the Maryland Commission on Oversight of Public-Private Partnerships*, MAPA Group, September 28, 2011, p. 9, http://www. docstoc.com/docs/99871353/Briefing-by-MichaelA-Peck-to-the-Maryland-Commission-on.

[93] U.S. Congress, House Committee on Ways and Means, Subcommittee on Select Revenue Measures, *Written Comments for the Record: Hearing for Certain Expiring Tax Credits*, 112th Cong., 2nd sess., April 26, 2012, p. 3, http://waysandmeans.house.gov/uploadedfiles /vestas_american_wind_tech._inc._fc42612.pdf.

[94] Ryan Wiser and Mark Bolinger, *2011 Wind Technologies Market Report*, Lawrence Berkeley National Laboratory, August 2012, p. 24, http://www1.eere.energy.gov/wind /pdfs/2011_wind_technologies_market_report.pdf.

[95] National Export Initiative, *Renewable Energy and Energy Efficiency Export Initiative*, December 2010, http://export.gov/reee/eg_main_023036.asp.

[96] GTIS, Global Trade Atlas database (accessed December 11, 2012).

[97] Ryan Wiser and Mark Bolinger, *2011 Wind Technologies Market Report*, Lawrence Berkeley National Laboratory, August 2012, p. 17, http://www1.eere.energy.gov/wind /pdfs /2011_wind_technologies_market_report.pdf.

[98] Andrew S. David and Dennis Fravel, *U.S. Wind Turbine Export Opportunities in Canada and Latin America*, United States International Trade Commission, No ID—032, July 2012, p. 11-26, http://www.usitc.gov/publications/332/ working_papers/ID-032_final.pdf.

[99] Ibid., pp. 23-32.

[100] World Trade Organization, Tariff Analysis Online, http://www.wto.org/english/tratop_e /tariffs_e/tariff_data_e.htm.

[101] Clean energy policies in China, Japan, and South Korea are detailed in a November 2009 study by the Breakthrough Institute and the Information Technology & Innovation Foundation, *"Rising Tigers Sleeping Giant: Asian Nations Set to Dominate the Clean Energy Race by Out-Investing the United States"* http://thebreakthrough.org/blog/ Rising_Tigers.pdf.

[102] More information about the Export-Import Bank's Environmental Exports Program can be accessed at http://www.exim.gov/products/policies/environment/success.cfm.

[103] Export-Import Bank of the United States, "Clipper Windpower Transaction is Named Ex-Im Bank Deal of the Year," press release, March 11, 2010, http://www.exim.gov/press release.cfm/4EB6A01A-B9E1-FABFD9409670AEB9668D/.

[104] Export-Import Bank of the United States, "Vermont Manufacturer Makes Largest U.S. Export Of Distributed Wind Turbines, Ex-Im Bank Guarantees Financing," press release, May 26, 2011, http://www.exim.gov/newsandevents/ releases/2011/vermont-manufacturer-makes-largest-u-s-export-of-distributed-wind-turbines-ex-im-bank-guaranteesfinancing.cfm.

[105] Craig O'Connor, *Financing Cleantech Exports: The Role of Ex-Im Bank*, Export-Import Bank of the United States, December 14, 2011, p. 15. The Ex-Im Bank reported that the wind turbines would be manufactured at its facility in Pennsylvania and generators would be supplied by ABB Power T&D Company (Bland, VA), blades by LM Glassfiber Inc. (Grand Forks, ND), and associated equipment and services from other U.S. suppliers.

[106] Export-Import Bank of the United States, "Ex-Im Approves $32.1 Million in Financing For Export of U.S. Wind Blades to Brazil," press release, August 3, 2012, http://www.exim.gov/ newsandevents/releases/2012/ex-im-approves32-1-million-in-financing-for-export-of-u-s-wind-blades-to-brazil.cfm.

[107] KPMG International, *Taxes and Incentives for Renewable Energy*, June 2012, http://www.kpmg.com/Global/en/ IssuesAndInsights/ArticlesPublications/Documents/taxes-incentives-renewable-energy-2012v3.pdf.

[108] Siemens, *Current Trends in Renewable Energy Markets*, p. 3, April 2012, http://www.iea.org/media/weowebsite/ebc/meetings/ebcmeeting-12-13june2012/Siemens TrendsinRenewableEnergyMarkets.pdf.

[109] A feed-in tariff, or FIT, is a renewable energy policy that typically offers a guarantee of payments to project owners for the total amount of renewable energy they produce; access to the grid; and stable, long-term contracts (15-20 years). For more information see workshop presentation, Renewable Energy Feed-in Tariffs: An Analytical View, by Toby Couture, May 28, 2009. http://www.energy.ca.gov/2009_energypolicy/documents/2009-05-28_workshop/ presentations/01_Couture_Feed-in_Tariff_Wkshop_May_28_09.pdf.

[110] European Wind Energy Association, *Support Schemes for Renewable Energy, A Comparative Analysis of Payment Mechanisms in the EU*, 2002, p. 31, http://www.ewea.org/fileadmin /ewea_documents/documents/projects/rexpansion/ 050620_ewea_report.pdf.

[111] At least three studies have raised questions about the costs associated with Europe's support of its renewable energy sectors. A report by a Spanish academician, Dr. Gabriel Calzada, *Study of the Effects on Employment of Public Aid to Renewable Energy Sources*, argued that Spain's policies were an economic failure and cost many jobs. Another report by a Danish think tank, CEPOS, *Wind Energy: The Cost for Denmark*, also pointed to the costs of subsidizing Denmark's wind power industry. A third report by the German think tank, Rheinisch-Westfälisches Institut for Wirtschaftsforschung (RWI), *Economic Impacts from the Promotion of Renewable Energies: The German Experience*, argues that aid by the German government for wind power is now three times the cost of conventional electricity.

[112] Ben Backwell, "Subsidies to be Cut for Spain's Wind and Thermal Solar Sectors," *Recharge News*, July 5, 2010.

[113] For a detailed discussion of China's green energy policies, see CRS Report R41748, *China and the United States—A Comparison of Green Energy Programs and Policies*, by Richard J. Campbell.

[114] GWEC, *Global Wind Report Annual Market Update 2010*, April 2011, pp. 30-33, http://www.gwec.net/index.php? id=180.

[115] Eric Martinot, *Renewable Power for China: Past, President and Future*, 2010, p. 6, http://www.martinot.info/ Martinot_FEP4_prepub.pdf.

[116] AWEA, *Production Tax Credit*, What is the Production Tax Credit?, p. 1, http://www.awea.org/_cs_upload/issues/ federal_policy/7785_1.pdf.

[117] The PTC expired for the first time on July 1, 1999. In December 1999, the PTC extension was included as part of the Ticket to Work and Work Incentives Improvement Act of 1999, P.L. 106-170. The PTC ended again on January 1, 2002, but was extended through December 31, 2003, by the Job Creation and Worker Assistance Act of 2002, P.L. 107-147. Then at year-end 2003, it expired again and was renewed nine months later by the Working Families Tax Relief Act of 2004, P.L. 108-311, which included a PTC sunset date of December 31, 2005. In August 2005, a PTC extension was included in the Energy Policy Act, P.L. 109-58, which stretched the credit through December 31, 2007. In December 2006, Congress continued the PTC for an additional year ending on December 31, 2008, as part of the Tax Relief and Health Care Act of 2006, P.L. 109-432. Thereafter, the wind PTC was extended through the end of 2009 in the Emergency Economic Stabilization Act of 2008, P.L. 110-343. A provision in the American Recovery and Reinvestment Act of 2009, P.L. 111-5, extended the deadline to December 31, 2012, and in January 2013, Congress enacted the American Taxpayer Relief Act of 2012 (P.L. 112-240), which prolonged the PTC for one more year covering those wind projects that begin construction by year-end 2013.

[118] U.S. Congress, House Committee on Science, Space, and Technology, Subcommittee on Investigations and Oversight, *Impact of Tax Policies on the Commercial Application of Renewable Energy Technology*, 112th Cong., 2nd sess., April 19, 2012, p. 10, http://science.house.gov/sites/republicans.science.house.gov/files/documents/hearings/ HHRG-112-SY21-WState-LLinowes-20120419.pdf.

[119] American Taxpayer Relief Act of 2012 (ATRA; P.L. 112-240), January 2, 2013, http://www.gpo.gov/fdsys/pkg/ BILLS-112hr8enr/pdf/BILLS-112hr8enr.pdf.

[120] AWEA, *Federal Policy*, https://www.awea.org/issues/federal_policy/index.cfm.

[121] Internal Revenue Service Notice 2009-52, Election of Investment Tax Credit, Coordination with Department of Treasury Grants for Specified Energy Property in Lieu of Tax Credits, http://www.irs.gov/pub/irs-drop/n-09-52.pdf.

[122] In a December 12, 2012, letter to Congress, AWEA proposed a six-year reduction of the PTC between 2014 and 2018, with the tax credit starting at 100% of the current 2.2 cents a kilowatt-hour for projects started in 2013, followed by 90%, 80%, 70%, 60%, and then 60% of the current project level for projects placed in service from 2014 through 2018, with no PTC in 2019 or thereafter. AWEA, *Phase-Out of Wind Energy Production Tax Credit Would Enable U.S. Industry To Become Full Cost-Competitive*, December 12, 2012, http://www.awea.org/newsroom/pressreleases/ Analysisonphaseout.cfm.

[123] Letter from Governor's Wind Energy Coalition to Congress, November 15, 2011, http://governorswindenergycoalition.org/wp-content/uploads/2011/03/GWC-PTC-Letter-Final2-11-15-11.pdfhttp:// www.governorswindenergycoalition.org/assets/files/President Obama Wind Energy Letter %28July 24, 2011%29.pdf.

[124] For more information see White House, *Fact Sheet: $2.3 Billion in New Clean Energy Manufacturing Tax Credits*, January 8, 2010, http://www.whitehouse.gov/the-press-office/fact-sheet-23-billion-new-clean-energy-manufacturingtax-credits.

[125] "SEAM Act Will Build U.S. Wind Supply Chain Says Industry Group," *Industry Week*, May 12, 2010. http://www.industryweek.com/articles/seam_act__will_help_build_u-s-_wind_supply_chain_says_industry_group_21813.aspx?SectionID=2.

[126] White House, *The Recovery Act: Transforming the American Economy Through Innovation, Promoting Clean, Renewable Energy: Investments in Wind and Solar*, http://www.whitehouse.gov/recovery/innovations/clean-renewableenergy.

[127] The Tax Relief, Unemployment Insurance Reauthorization, and Job Creation Act of 2010 (P.L. 111-312) was signed by President Obama on December 17, 2010.

[128] Ryan Wiser and Mark Bolinger, *2011 Wind Technologies Market Report*, Lawrence Berkeley National Laboratory, August 2012, p. 57.

[129] Criteria for the start of construction are detailed in a U.S. Treasury guidance document, which can found on the Treasury Department's 1603 Grant Program website at http://www.treasury.gov/initiatives/recovery/Pages/1603.aspx.

[130] Mark Bolinger, Ryan Wiser, and Naim Darghouth, *Preliminary Evaluation of the Impact of the Section 1603 Treasury Grant Program on Renewable Energy Deployment in 2009*, Ernest Orlando Lawrence Berkeley National Laboratory, April 2010, p. ii, http://eetd.lbl.gov/ea/emp/reports/lbnl-3188e.pdf.

[131] Solar generation or solar manufacturing comprised the overwhelming majority (85%) of the 1705 loan guarantee projects funded at $15.9 billion by the Department of Energy. A list of the 1705 Loan Program projects can be found at https://lpo.energy.gov/?page_id=45.

[132] Department of Energy, "DOE LPO Finalizes Deal on the World's Largest Wind Project to Date," press release, December 17, 2010, https://lpo.energy.gov/?p=1955.

[133] Nordic Windpower had received a conditional commitment of $16 million in 2009 for a wind turbine manufacturing project in Kansas City, MO, but abandoned its application in 2011. The company filed for Chapter 7 bankruptcy protection in October 2012.

[134] CRS Report R42576, *U.S. Renewable Electricity: How Does the Production Tax Credit (PTC) Impact Wind Markets?*, by Phillip Brown.

[135] U.S. Department of Energy, *2010 Wind Technologies Market Report*, August 2012, p. 58.

[136] See CRS Report R41493, *Options for a Federal Renewable Electricity Standard*, by Richard J. Campbell.

[137] CRS Report R42818, *U.S. Renewable Electricity: How Does Wind Generation Impact Competitive Power Markets?*, by Phillip Brown.

INDEX

C